Sarah.
I know you
love food issues
George Ke

D0796690

FOOD
SOVEREIGNTY

Reconnecting Food, Nature & Community

Edited by Hannah Wittman,
Annette Aurélie Desmarais & Nettie Wiebe

**Pambazuka
Press**

FERNWOOD
PUBLISHING

FOODFIRST
B O O K S

BOOKS *for* CHANGE
Dedicated to Development
International Publishing House

This edition first published in 2011 in Europe, Africa and Asia by Pambazuka Press, an imprint of Fahamu
Cape Town, Dakar, Nairobi and Oxford
2nd floor, 51 Cornmarket Street, Oxford OX1 3HA, UK
www.pambazukapress.org
Pambazuka ISBN 978-0-85749-029-2 (pb)

Published in Canada by Fernwood Publishing
32 Oceanvista Lane, Black Point, Nova Scotia, B0J 1B0
and 748 Broadway Avenue, Winnipeg, MB R3G 0X3
www.fernwoodpublishing.ca
Fernwood Publishing ISBN 978-1-55266-374-5 (pb) ISBN 978-1-55266-390-5 (hb)

Published in the United States by Food First Books
398 60th Street, Oakland, CA 94618-1212
510-654-4400, www.foodfirst.org
Food First ISBN 978-0-935028-37-9

Published in India by Books *for* Change
139 Richmond Road, Bangalore 560 025, India
www.booksforchange.info
Books for Change ISBN 978-81-8291-108-6

British Library Cataloguing in Publication Data
A catalogue record for this book is available from the British Library

Printed by National Printing Press, Bangalore, India

Contents

Acknowledgements

Our first and deepest debt is to the millions of people around the world who engage in the valiant, multifaceted struggle for food sovereignty — most especially the peasants and small-scale farmers of La Vía Campesina. This book is humbly dedicated to you.

Many of the contributors to this book are authors deeply involved in public food sovereignty debates who braved winter travel for a two-day workshop in Saskatoon, Saskatchewan, Canada to exchange ideas and knowledge about food sovereignty. The workshop was made possible through grants from the Social Sciences and Humanities Research Council of Canada, the University of Regina, and Simon Fraser University. It was generously hosted by St. Andrews College and the National Farmers' Union, with special thanks to Martha Robbins for logistical support.

We would like to thank Jim Handy for his assistance in editing some chapters, as well as the graduate research assistants who lightened and improved our work: Christina Bielek, Chris Hergesheimer and Rachel Elfenbein at Simon Fraser University; Yolanda Hansen and Naomi Beingessner at the University of Regina; and Terran Giacomini at the University of Guelph.

During the course of delivering this book, which represents the gestation of ideas, struggles, theories and critiques, the literal delivery of baby Eli added anticipation, joy — and a good deal of work for Hannah. The birth of Eli, like that of every new child, highlights the crucial importance and responsibility of ensuring sustainable, regenerative food systems into the future.

Finally, thank you to Wayne Antony for his interest and enthusiasm and to the staff of Fernwood Publishing — Brenda Conroy, Beverley Rach and Debbie Mathers — for seeing this collective project through to final production.

Acronyms

ADM	Archer Daniels Midland
AGRA	Rockefeller/Gates Alliance for a Green Revolution in Africa
CESCR	International Covenant on Economic, Social and Cultural Rights
CGIAR	Consultative Group on International Agricultural Research
CSO	civil society organizations
FAO	Food and Agriculture Organization of the United Nations
FOSS	free and open software
FSPI	Federation of Indonesian Peasant Union
GM	genetically modified
GMO	genetically modified organism
GPLPG	general public license of plant germplasm
GRAIN	Genetic Resources Action International
IAASTD	International Assessment of Agricultural Knowledge, Science and Technology for Development
ICC	International Coordinating Commission (of La Vía Campesina)
IFC	International Finance Corporation (of the Inter-American Development Bank and the World Bank)
IFI	international financial institution
IMF	International Monetary Fund
INIREB	Mexican National Research Institute on Biotic Resources
IPC	International Planning Committee for Food Sovereignty
IPR	intellectual property rights
LDC	least developed countries
MCAC	Moviemento Campesino a Campesino
MST	Brazilian Landless Workers' Movement
NAFTA	North American Free Trade Agreement
NGO	non-governmental organization
NSCM	National Seed Company of Malawi
OECD	Organization for Economic Cooperation and Development
OPV	open pollinated varieties (of seed)
OWINFS	Our World Is Not for Sale
RFA	Renewable Fuels Association
ROPPA	Network of Farmers' and Agricultural Producers' Organizations of West Africa
TAM	transnational agrarian movement
TRIPS	Trade-Related Aspects of Intellectual Property Rights
UDHR	Universal Declaration of Human Rights
USAID	United Sates Agency for International Development
USDA	United States Department of Agriculture
WTO	World Trade Organization

Editors

Annette Aurélie Desmarais was a cattle and grain farmer in Canada for fourteen years. She is now an associate professor of international studies at the University of Regina. She has worked as technical support with La Vía Campesina since its inception in 1993 and recently published *La Vía Campesina: Globalization and the Power of Peasants*. Annette's research focuses on rural social movements, food sovereignty, gender and rural development.

Nettie Wiebe is an organic farmer and professor of ethics at St. Andrew's College, University of Saskatchewan. She was women's president of the National Farmers' Union and then served four years as the president of the NFU — the first and only woman to have led a national farmers' organization in Canada — as well as serving as a member of La Vía Campesina's International Coordinating Commission (ICC). She is actively involved in local and national politics in Canada, and her research focuses on agrarian feminism and the intersection of environmental, agricultural and women's issues in rural communities.

Hannah Wittman is an assistant professor of sociology and Latin American studies at Simon Fraser University. She conducts collaborative research on local food systems, farmer networks and agrarian citizenship in British Columbia, and in Latin America with Brazil's Landless Rural Workers Movement (MST) and La Vía Campesina. Her research interests are in environmental sociology, agrarian citizenship and agrarian social movements.

Contributors

Miguel A. Altieri is a professor of agroecology at UC Berkeley and has served as scientific advisor to the Latin American Consortium on Agroecology and Development. He was the general coordinator for the UNDP's Sustainable Agriculture Networking and Extension Programme and chair of the NGO committee of the Consultative Group on International Agriculture Research. Currently he is advisor to the FAO Globally Ingenious Agricultural Heritage Systems, director of the U.S.-Brazil Consortium on Agroecology and Sustainable Rural Development and general coordinator of the Latin American Scientific Society of Agroecology. He is the author of more than two hundred publications and numerous books, including *Agroecology; Pest Management in Agroecosystems* and *Agroecology and the Search for a Truly Sustainable Agriculture*.

Mara Baviera is a researcher in development studies. She has a political science degree from the University of the Philippines and worked as research assistant for

Walden Bello on the book *Food Wars*. She has also worked with the Third World Studies Center assisting in research projects on human security, Philippine trade negotiations in the World Trade Organization, technocracy and economic decision-making in the Philippines and multilateral and bilateral lending institutions' funding trends to developing countries.

Walden Bello is a professor at the University of the Philippines at Diliman, a fellow at the Transnational Institute, executive director of Focus on the Global South and a recently elected Member of Parliament in the Philippines He is the author of numerous books including *The Food Wars, Dilemmas of Domination, The Anti-Development State* and *Deglobalisation*. Walden has been on the front lines of activism and research on globalization and transnational peasant movements for the last three decades. In 2003 he received the Right Livelihood Award, also known as the Alternative Nobel Prize, "for outstanding efforts in educating civil society about the effects of corporate globalization, and how alternatives to it can be implemented."

Rachel Bezner Kerr is an assistant professor in the Department of Geography at the University of Western Ontario. She is the research coordinator for the Soils, Food and Healthy Communities Project based in Malawi and has worked with farmer communities on seed-related research in Northern Malawi for over ten years. Within the field of international development, she conducts research on social and environmental inequalities and community-based, participatory research related to agriculture, health and nutrition.

Saturnino Borras Jr. is Canada Research Chair in International Development Studies at Saint Mary's University. He was part of the founding group of La Vía Campesina and was a member of its International Coordinating Commission from 1993 to 1996. He is the author of *Pro-Poor Land Reform, Competing Views and Strategies on Agrarian Reform* (Volumes 1 and 2), and he is co-editor of *Critical Perspectives in Rural Development Studies* and *Transnational Agrarian Movements Confronting Globalization*. His research interests include land policies, (trans) national agrarian movements, landgrabbing and agrofuels.

Madeleine Fairbairn is a PhD candidate in the departments of Sociology and Community and Environmental Sociology at the University of Wisconsin-Madison. Her master's thesis examined the framing of food sovereignty from a historical perspective. Her research interests include alternative agrifood systems, food politics, economic development and social justice.

Carla Fehr grew up on a farm in Saskatchewan, Canada, and has been actively involved in food and agricultural politics at the local level. She completed degrees in international studies at the University of Saskatchewan and history at the University of Saskatchewan. Her master's thesis, which examined the *Economist* newspaper's views on agricultural change from 1843 to 1863, is in the process of being revised into a monograph.

Jennifer C. Franco is a researcher at the Transnational Institute (TNI). Her research interests include rural democratization, human rights, land reform, agrarian movements, socio-political dimensions of law and the politics of agrofuels. She is the author of *Elections and Democratization in the Philippines* and co-editor of *On Just Grounds* and a forthcoming book *Law and the Rural Poor*.

Jim Handy is a professor of history and former director of the International Studies Program at the University of Saskatchewan. He has expertise in the history of agrarian social change, rural conflict and agrarian reform. He is the author of *Gift of the Devil: A History of Guatemala* (1984), *Revolution in the Countryside: Rural Conflict and Agrarian Reform in Guatemala, 1944–1954* (1994) and is currently working on *The Menace of Progress*.

Eric Holt-Giménez is the executive director of Food First/Institute for Food and Development Policy in Oakland. He has also worked as a lecturer in international development and agroecology at the University of California and Boston University. For three decades, Eric helped organize and train farm leaders and was a consultant in agroecology in Latin America. His books include *Campesino a Campesino* and *Food Rebellions! Crisis and the Hunger for Justice* (co-authored with Raj Patel and Annie Shattuck).

Jack Kloppenburg is a professor of rural sociology at the University of Wisconsin-Madison. His groundbreaking book first published in 1988, entitled *First The Seed*, was revised and reissued in 2004. He continues to do research on the global access to and control over genetic resources, and he publishes widely on in the area of food, culture and society.

Itelvina Masioli is a member of the National Coordinating Council, Landless Rural Workers Movement of Brazil (MST) and of the International Coordinating Commission of La Vía Campesina. Raised on a small family farm in Goais, Brazil, Itelvina has been active in integrating women's issues into the rural land struggle in Brazil for over twenty years and has taken a leading role in building coalitions of peasant and rural women's movements internationally. She is currently involved in educational programs at the MST's flagship school, Florestan Fernandes National University in São Paulo, Brazil.

Philip McMichael is a professor of development sociology at Cornell University. His book, *Development and Social Change* is now in its fourth edition. He also spearheaded research efforts at the International Sociological Association to integrate the study of global agrarian restructuring with analyses of peasant-movement mobilization and resistance. He recently worked with UNRISD and La Vía Campesina on the food crisis and agroecological alternatives. He has published extensively on agrarian social movements, food regimes and agrarian transformation.

Paul Nicholson was a dairy farmer for twenty years. He is the former president of the Basque Ranchers and Farmers' Union (*Euskal Herriko Nekazarien Elkartea*), and was a member of the board of the European Farmers' Coordination, which has since become the European Coordination Vía Campesina. He was a founding member of La Vía Campesina and served as a member of its International Coordinating Commission from 1996 to 2008. During that time he led peasant mobilizations at the Ministerial Conferences of the World Trade Organization held at Seattle, Geneva, Cancun and Hongkong. Paul is also an active member of a small cooperative that processes local food for local markets in the Basque Country.

Raj Patel has been conducting research for the past decade on the intersection of food, globalization, development policy, human rights and poor peoples' organizing, particularly in Africa. He is a visiting scholar at the Center for African Studies and the University of California at Berkeley, and an honorary research fellow in the School of Development Studies at the University of Kwazulu-Natal, South Africa. He is the author of *Stuffed and Starved* and a co-author (with Eric Holt-Gimenez and Annie Shattuck) of *Food Rebellions*. He has also written on political economy, food and society as well as gender relations.

Annie Shattuck is a policy analyst at the Institute for Food and Development Policy. Trained in biology and agroecology, she has worked in participatory action research, rural development and ecology research in the U.S. and Latin America. She is co-author of the book *Food Rebellions!* (with Eric Holt-Giménez and Raj Patel), which examines the root causes of the global food crisis and the grassroots solutions to hunger springing up around the world. Her writing has appeared in many popular print and online media as well as in academic venues.

1

The Origins & Potential
of Food Sovereignty

Hannah Wittman, Annette Desmarais & Nettie Wiebe

The global food crisis of 2007–08, marked by skyrocketing food prices, urban food riots and the continued displacement of the rural poor, was a clear indication that the dominant model of agricultural development has not succeeded in eradicating poverty or world hunger. In desperation, in Haiti, Bangladesh, Egypt, West and Central Africa and countless other locations, hundreds of thousands of people took to the streets demanding affordable food. Behind these highly visible events lurks the very real and ongoing human suffering caused by the lack of that key necessity for all human life — food. The stunted growth and high mortality rates of hungry children and the ill health and lost potential of malnourished adults are clear and tragic results of the chronic food shortages suffered by an increasing number of people. A growing number of households and communities fear for tomorrow's meals, even though there may be enough food for today. And even for those of us whose cupboards are well stocked and who have adequate incomes to pay our grocery bills, there are grounds for unease about the content, safety and origins of our food and the long-term sustainability of our food system. Hence, the security, cost, safety and nutrition of food and the future of food production itself are everyone's concern. While the sudden spike in prices sparked the headlines during the 2007–2008 food crisis, the problems in the global food system are complex and deep-seated. The food system's vulnerabilities, from climate change to loss of biodiversity to security of supplies, are becoming more apparent. The global food crisis is deepening. What are the possible solutions to this crisis?

Some proponents of neoliberal globalization would have us believe that the crisis is the result of shortages and market failures. They assure us that the best way to keep up with a growing population is to prevent national governments from intervening in the market, focus on scientific high-tech approaches, increase production with the adoption of genetically modified seeds (GMOs) and further liberalize agriculture and food. But despite having powerful advocates and enforcers, such as the World Bank, the International Monetary Fund (IMF) and the World Trade Organization (WTO) on side, these solutions reveal a spectacular failure when it comes to reducing poverty and eradicating hunger. The most recent figures from the Food and Agriculture Organization (FAO) of the United Nations indicate that the ranks of the hungry are continuing to swell and now encompass more than one billion people, an increase of over 25 percent in the number of people without enough food since the mid 1990s (FAO 1999, 2009), when the neoliberal development project was in a phase of full implementation.

As an alternative to the neoliberal model, peasants, small-scale farmers, farm workers and indigenous communities organized in the transnational agrarian movement La Vía Campesina (2008a) argue that the current, and linked, food, economic and environmental crises are in fact the direct result of decades of destructive economic policies based on the globalization of a neoliberal, industrial, capital-intensive and corporate-led model of agriculture. La Vía Campesina, formed in 1993 and now representing 148 organizations from sixty-nine countries, has become one of the strongest voices of radical opposition to the globalization of an industrial and neoliberal model of agriculture, claiming that "the time for food sovereignty has come."

Peasant movements, urban-based social movements, non-governmental organizations (NGOs) and indigenous peoples have been instrumental in putting food sovereignty on the agenda, and consequently, they have succeeded in shifting the terms of the debate around food, agriculture and rural development at the local, national and international levels. Because food sovereignty aims to transform dominant forces, including those related to politics, economics, gender, the environment and social organization, there will, no doubt, be a long and hard struggle to see food sovereignty become the standard model for food production and rural development. This book contributes to this struggle by engaging in a conversation that identifies and expands the meanings, understandings and implications of food sovereignty in an international context.

Initiating the Food Sovereignty Concept

Food sovereignty as a concept evolved from the experience of, and critical analysis by, farming peoples, those most immediately affected by changes in national and international agricultural policy introduced throughout the 1980s and early 1990s. The results of the inclusion of agriculture in the General Agreement on Tariffs and Trade (GATT) negotiations, articulated in the WTO, brought into sharp relief communities' widespread loss of control over food markets, environments, land and rural cultures. The term "food sovereignty" was coined to recognize the political and economic power dimension inherent in the food and agriculture debate and to take a pro-active stance by naming it. Food sovereignty, broadly defined as the right of nations and peoples to control their own food systems, including their own markets, production modes, food cultures and environments, has emerged as a critical alternative to the dominant neoliberal model for agriculture and trade.

La Vía Campesina (1996a) first discussed food sovereignty at its Second International Conference, held on April 18–21, 1996, in Tlaxcala, Mexico. Peasant and farm leaders who gathered there no longer saw potential in the concept of "food security" to ensure local access to culturally appropriate and nutritious food. In common usage, food security describes "a situation that exists when all people, at all times, have physical, social and economic access to sufficient, safe and nutritious food that meets their dietary needs and food preferences for an active and healthy life" (FAO 2001). This definition invites an interpretation towards food related policies that emphasizes maximizing food production and enhancing food access opportunities, without particular attention to how, where and by whom food is produced. This common definition also is uncritical of current patterns of food consumption and distribution.

Governments and agri-business corporations have pursued food security by promoting increased agricultural trade liberalization and the concentration of food production in the hands of fewer, and larger, agri-business corporations. Excess production is off-loaded through "dumping," an international trade strategy that places food in targeted export markets at prices below the cost of production. This practice has had devastating effects on domestic agricultural systems, which cannot compete with the influx of subsidized commodities saturating local markets. International aid agencies that subscribe to the view that food insecurity is primarily the result of a lack of supply have also opted for variations of this "just produce and/or import more food from somewhere" strategy.

These contemporary policies aimed at food security offer no real possibility for changing the existing, inequitable, social, political and economic structures and policies that peasant movements believe are the very causes of the social and environmental destruction in the countryside in both the North and the South. To counter these structures and policies, La Vía Campesina (1996a) proposed a radical alternative, one "directly linked to democracy and justice," that put the control of productive resources (land, water, seeds and natural resources) in the hands of those who produce food. The Tlaxcala Conference defined eleven principles of food sovereignty, all of which were then integrated into La Vía Campesina's (1996b) Position on Food Sovereignty, presented at the World Food Summit in Rome in November 1996 (see Appendix 1).

Subsequently, La Vía Campesina worked with other organizations and civil-society actors to further elaborate the food sovereignty framework. Here, two international civil-society events, among others, proved significant: the World Forum on Food Sovereignty held in Cuba (2001) and the NGO/CSO Forum on Food Sovereignty (2002) held in Rome in conjunction with the World Food Summit: Five Years Later. Perhaps most importantly, the international coalition Our World Is Not for Sale (OWINFS) helped form an international food sovereignty network of social movements, research institutions and NGOs to collectively develop the People's Food Sovereignty Statement (People's Food Sovereignty Network 2001), which includes specific international mechanisms to ensure food sovereignty (see Appendix 2). This statement, along with La Vía Campesina's 1996 World Food Summit document, is the most often cited international declaration of food sovereignty and was developed and signed by many of the same organizations that now form part of the International Planning Committee for Food Sovereignty (IPC). These more recent documents reflect many aspects of La Vía Campesina's original position at the 1996 World Food Summit.

The Scope of Food Sovereignty

Can food sovereignty address the multiple crises affecting food and agriculture around the world? Can food sovereignty ensure sufficient, healthy food for everyone and provide livelihoods for peasant farmers while redressing decades of environmental degradation caused by industrial agriculture? What would change if the concept of food sovereignty were widely adopted?

The theory and practice of food sovereignty has the potential to foster dramatic and widespread change in agricultural, political and social systems related to food by posing a radical challenge to the agro-industry model of food production. The transformation envisioned entails a changing relationship to food resulting from an integrated, democratized, localized food production model. It also entails a fundamental shift in values expressed in changed social and political relations. At an international workshop on food sovereignty, Jim Handy (2007), a professor of history at the University of Saskatchewan, summarized the revolutionary implications of the seemingly simple idea of democratizing the food system:

> I would like to express my sense of awe at the enormity of the change that is envisioned through the concept of food sovereignty. Food sovereignty challenges not just a particular development model, doesn't just challenge a particularly abhorrent form of neo-liberalism, doesn't just suggest a new set of rights. Rather, it envisions fundamental changes in the basis of modern society. Modern society was based on a set of exclusions and enclosures that were fundamental to the emergence and strengthening of capitalism. Those exclusions were felt primarily in the countryside and primarily in agriculture. Capitalism was dedicated to divorcing producers from any right over the goods they produced and encasing those goods in ever larger, ever more disconnected, ever more monopolized, and ever more destructive markets. Food sovereignty challenges all of that because it demands that we rethink what was at the very centre of this transition; it demands that we treat food not simply as a good, access to which and the production of which is determined by the market, it demands that we recognize the social connections inherent in producing food, consuming food, and sharing food. In the process it will change everything.

Certainly, ideas about food sovereignty force us to rethink our relationships with food, agriculture and the environment. But, perhaps the most revolutionary aspect of food sovereignty is that it forces us to rethink our relationships with one another. The magnitude of this transformation hit home in a powerful way when, during its Fifth International Conference, La Vía Campesina launched a campaign with the slogan, "Food sovereignty means stopping violence against women."[1] Because women play a key role in food production and procurement, food preparation, family food security and food culture, the social and political transformation embedded in the food sovereignty concept specifically entails changed gender relations. Food sovereignty for communities and peoples cannot be achieved without ensuring equality, respect and freedom from violence for women. As the Declaration of Maputo stated: "If we do not eradicate violence towards women within our movement, we will not advance in our struggles, and if we do not create new gender relations, we will not be able to build a new society" (La Vía Campesina 2008b).

The theoretical context of the food sovereignty struggle is framed by the changing relationship to food imposed by the industrialization of production and the globalization of agricultural trade. The globalized food system distances eaters from the people who produce food and from the places where food is produced — literally and conceptually. The more industrialized, processed and distant food is, the less connected to and knowledgeable about it the consumer becomes. This paucity of knowledge changes our relationship to our meals, stripping meaning, cultural significance and even appreciation from our daily food experiences. But it also undermines our capacity for making decisions about this key determinant of our lives and our economies. Hence the inextricable connection between food, culture and democracy.

As an alternative approach to rural development, the concept of food sovereignty is not limited to how, where and by whom food is produced but is integrally linked to other issues facing rural and global society. In this book, we discuss many of these issues, including the link between the right to food and other human rights; the exploration of alternative notions of citizenship to include participatory-democratic structures and practices related to rural and urban food production and distribution; and the relationship between food production, resource redistribution and environmental and social wellbeing.

Broadening the Struggle for Food Sovereignty

The idea of food sovereignty has gained significant momentum as numerous local, national and international social movements and NGOs have embraced it in efforts to shift agriculture and food policy (NOUMINREN 2006, International Workshop on the Review of the Agreement on Agriculture 2003, Nyéléni 2007). Some of these initiatives involve recognizing the specific implications of food sovereignty for specific local and regional populations, as in the case of the European Platform for Food Sovereignty, Task Force Food Sovereignty in the Philippines, the People's Coalition on Food Sovereignty and the People's Caravan for Food Sovereignty, which involves a coalition of Asian agricultural and peasant movements. For instance, the movement for indigenous food sovereignty in Western

Canada, involving the rights of traditional populations to "hunt, gather, fish, grow, and eat" (Morrison 2008), must grapple with the competing demands of local agricultural populations to expand their own productive capacity. Similarly, urban food sovereignty networks seek ways to protect and link local food systems to urban consumers, who increasingly recognize and demand access to local food.

The food sovereignty movement also seeks to influence policy change at an international level through global coalition building. The People's Food Sovereignty Network, mentioned earlier, is a powerful example of this. The International NGO/ CSO Planning Committee for Food Sovereignty (IPC) — a global network bringing together representatives of indigenous peoples, fisherfolk, farmers/peasants, youth, women and NGOs from many regions of the world — plays the key role of global coordination and communication. The IPC emerged in 2000 as a coalition of fifty-two civil-society organizations (CSOs), including La Vía Campesina, to plan a collective approach to the 2002 World Food Summit: Five Years Later. It works to develop common positions within the network, which are then presented to international institutions and key meetings.

This global network was instrumental in organizing the parallel NGO/CSO Forum for Food Sovereignty in the context of the World Food Summit: Five Years Later and is responsible for carrying out the action agenda adopted by this forum. In 2003, the IPC reached agreement with the director general of the FAO to act as the principal civil-society interlocutor for follow up to the 2002 summit and developed a work plan to advance a civil-society agenda in four priority areas: the right to food and food sovereignty; access to, management of and local control of resources; small-scale, family-based, agroecological food production; and trade and food sovereignty. Subsequently, the IPC facilitated the organization of civil-society consultations with the FAO and ensured representation of farmer and peasant organizations to several FAO technical committees, including fisheries, commodity products, agriculture and world food security.

The IPC also helps to organize regional and international gatherings around food sovereignty, including an especially important event, the Nyéléni International Forum on Food Sovereignty, held in February 2007 in Nyéléni, Mali. Just over a decade after introducing food sovereignty in the international arena, La Vía Campesina worked as a member of the Nyéléni Forum steering committee, which included the Network of Farmers' and Producers' Organizations of West Africa, World Women's March, World Forum of Fish Harvesters and Fisherworkers, World Forum of Fisher Peoples, Friends of the Earth International and others involved in the IPC, to organize a global forum to deepen understandings of food sovereignty, enhance common actions and solidarity and develop strategies for implementing food sovereignty at the local and global levels. The event brought together five hundred representatives of social movements, peasant movements, pastoralists, indigenous peoples, fishers, migrant workers and NGOs from eighty countries based in the North and the Global South, who vowed to continue to work together to build alliances at all levels and in so doing strengthen the global movement for food sovereignty (Nyéléni 2007).

By most accounts the Nyéléni forum was highly successful. First, it effectively moved food sovereignty beyond the producers' perspective and production, to include consumers' associations and consumption, something that La Vía Campesina was anxious to do. As Paul Nicholson (2007), leader with the Basque Country's peasant movement and former member of the International Coordinating Commission of La Vía Campesina, stated:

> What is motivating people to take on board food sovereignty? It is food insecurity, heating up of the planet, ecological crisis, longer food miles and the need for food quality and local economies. These are citizens' preoccupations, peoples' preoccupations. La Vía Campesina does not own food sovereignty. Food sovereignty was not designed as a concept only for farmers, but for people — this is why we call it peoples' food sovereignty. We see the need for a bottom up process to define alternative practices — an international space or platform for food sovereignty. We're talking about identifying allies, developing alliances with many movements of fisher folk, women, environmentalists and consumer associations, finding cohesion, gaining legitimacy, being aware of co-optation processes, the need to strengthen the urban-rural dialogue, to generate alternative technical models. And above all there is the issue of solidarity.

Second, the forum reached consensus on a vision of food sovereignty that sees food as being integral to local cultures, closes the gap between production and consumption, is based on local knowledge and seeks to democratize the food system. Third, the gathering provided a space where national and international coalitions were solidified. Finally, after Nyéléni, there was no doubt that we were now talking about a global food sovereignty movement that clearly understood the challenges ahead. As the Nyéléni documents stated,

> Food sovereignty is more than a right; in order to be able to apply policies that allow autonomy in food production it is necessary to have political conditions that exercise autonomy in all the territorial spaces: countries, regions, cities and rural communities. Food sovereignty is only possible if it takes place at the same time as political sovereignty of peoples. (Nyéléni 2007: 5)

The discourse of food sovereignty has thus entered the official international stage. In addition to the FAO's support for the IPC, reports submitted by the former Special Rapporteur on the Right to Food to the United Nations Commission on Human Rights (now replaced by the Council on Human Rights) advocate food sovereignty as the path to ensure peoples' human right to food and food security (Ziegler 2003, 2004). And, although he did not use the language of food sovereignty, the statement by Olivier De Schutter (2009), the current Special Rapporteur on the Right to Food, stressed the need to consider alternative sustainable models of agricultural development to ensure the full realization of the right to food. Key aspects of a sustainable model, he argued, are access to and secure tenure on land for the most vulnerable, regulation of transnational corporations and reorientation of national and international policies — all of which are components of food sovereignty.

All of this local, national and international grassroots activity and pressure for policy change has prompted a variety of responses from different levels of government and international bodies, some more pro-active than others. In recognizing the need for policies that support social, economic and environmental sustainability, numerous local mayors in several European nations have signed petitions endorsing a key element of food sovereignty: local production for local consumption. The Green Party in some European countries has held meetings on the subject to examine how it might help redefine European agricultural policy.

Several national governments have also integrated food sovereignty into their national constitutions and laws. For example, between 1999 and 2009, food sovereignty was included in national legislation promulgated by the governments of Venezuela, Mali, Bolivia, Ecuador, Nepal and Senegal. Certainly, there is the question of the extent to which these countries will succeed in further articulating laws and creating the necessary structures and mechanisms to implement the kind of genuine food sovereignty that will transform existing agriculture and food systems. For our purposes it is important to recognize that as a result of the strong mobilization and engagement of peasant organizations, NGOs and urban-based movements and the election of progressive political parties to power, these countries are in the process of creating, and in some cases already have opened, political spaces for debates within their borders about alternatives (Beauregard 2009).

Of course, social movements face some very real obstacles in their attempts to implement food sovereignty, as is clearly the case in Ecuador. With the election of the left-leaning President Rafael Correa in 2006, many social movements in the country saw potential for significant changes in agricultural policy. Consequently, three major rural movements — the Federación Nacional de Organizaciones Indígenas y Negras (National Federation of Peasant, Indigenous and Black People's Organizations), the Federación Nacional de Trabajadores Agroindustriales, Campesinos e Indígenas Libres del Ecuador (National Federation of Agro-Industry Workers, Peasants and Indigenous Peoples) and the Confederación Nacional Campesina-Eloy Alfaro (National Confederation of Peasants, CNC-Eloy Alfaro) — formed a coalition called La Mesa Agrária (Agricultural Roundtable) to put food sovereignty on the national agenda.[2] The coalition and other organizations, such as the Confederación de Nacionalidades Indígenas del Ecuador (Confederation of Indigenous Nationalities of Ecuador), mobilized sufficient support for food sovereignty that they subsequently worked with Ecuador's Constituent Assembly on drafting the new national constitution. In late September 2008, Ecuador approved its constitution, sections of which speak directly to elements of food sovereignty (Peña 2008). For example, the constitution bans the use of genetically modified seeds (except in the interests of national security), provides extension services and support for ecological agriculture and, as Article 281 states, "Food sovereignty constitutes an objective and strategic obligation from the State" (Government of Ecuador 2008).

With the constitution in hand, Ecuador's National Assembly began work on developing legislation to implement food sovereignty. After public consultations on February 17, 2009, the National Assembly passed its progressive Organic Law on the Food Sovereignty Regime. However, in March 2009, President Correa delivered a partial veto of the new law, citing concerns about the ban on GMOs, consequences of changes in land ownership structures and issues related to the production of agrofuels.[3] Newspapers reported that some social movements and government officials believe that the president acted under pressure from agri-business (El Universo 2009a). In a telling gesture, during the same month, the Government of Ecuador withdrew its support for Acción Ecológica (Environmental Action), a major environmental organization that had been working on food sovereignty (*El Universo* 2009b).

Moreover, although the new constitution contains groundbreaking articles supporting environmental sustainability, in January 2009, the National Assembly passed a new Mining Law to spur extraction in new areas by national and international companies. Many of Ecuador's social movements argue that this law undermines many of the constitution's guarantees, such as the rights to clean and safe water and a healthy environment — both important elements of food sovereignty. As Dosh and Kligerman (2009: 3) state, the new law jeopardizes those aspects of the constitution's articles that "ascribe to the environment itself the right to be respected, sustainably maintained, and regenerated." Without a doubt, we need to observe and analyze carefully what happens in Ecuador as those experiences hold important lessons for food sovereignty movements and national governments elsewhere.

Exploring Key Aspects of Food Sovereignty

The immediate and practical context that informs the discussion in this book is the literal displacement of millions of families from the land and their rural communities. Rural displacement and relocation is the other side of the large-scale, rapid urbanization evidenced by the exponential growth of cities over the last half century. Beginning in 2005, for the first time in history, less than half of the world population now lives and works in rural areas (UN 2007). Attendant on these massive movements of peoples is the social disruption and challenges of constructing human spaces in new places while coping with increased mobility and greater economic and political uncertainties. Food insecurity is among the most contentious and disruptive of these uncertainties.

In the first section of this book, we explore the genesis of the alternative model for food systems expressed by the language of food sovereignty. Resistance to the cultural, economic, ecological and social dislocations and destruction perpetrated by the current neoliberal, industrialized, corporate-led food regimes is more than just a sporadic or strategic "fight back" — although that remains an important element. The voices of leaders of La Vía Campesina articulate the praxis/theory interplay in describing and reflecting on their practical experiences of engaging in both resistance and in building alternatives. But this interplay necessarily takes place within an ideological framework that is itself contested by the new language and concepts. The movement to reframe thinking and shift dominant paradigms of food production and food culture emerges in opposition to an ideological history expressed by current food regimes. The origins of the food sovereignty discourse and movement, the analysis of the conceptual framework and food regimes challenged by food sovereignty and the first-hand experiences and reflections of leaders of the food sovereignty movement are all key entry points into the food sovereignty debate.

The hard reality about food is that, although its true value may reside in its nutritional and cultural benefits, it is largely and increasingly a priced commodity, subject to market conditions. The industrialization and capitalization of food production and the commodification of food have radically altered our relationships to food, land and place. The powerful economic forces unleashed by capitalism are examined from several vantage points in our second section. Beginning with a critical historical perspective and analysis of the economic genesis and outcomes of the current crisis not only offers a deeper understanding of the virulent economic forces undermining healthy food systems, it also points to places of resistance and reorganization. Diverting food into other uses such as agrofuels continues the further commodification, capitalization and corporatization of agricultural production. Far from being a rural development opportunity, the agrofuels "solution" represents an intensification of the economic pressures threatening food security and rural communities. Food sovereignty is thus a counterforce to the economic pressures and corporate values that the agrofuels strategy highlights.

The economic, social and political tensions of the current food system are accompanied and aug uented by increasingly obvious ecological stresses. Industrialized food producti on, characterized by intensification, chemical inputs, water and soil degradation, d eforestation and unsustainable resource exploitation, is causing major ecological dis uptions. We are all inescapably confronted by environmental problems that require fundamental changes of the food system. While much of the mainstream information about the food system focuses on the market components such as marketing, trade, packaging, nutrient and safety regulations and branding, building ecologically sustainable food systems requires fundamental changes of values and relationships. Gaining an understanding of these changes means more than simply decrying the environmental destruction caused by current practices; importantly, it entails building what is best called a new "agrarian citizenship." In section three, authors examine the ways in which contemporary peasant movements are revaluing the relationship between agriculture, land and the environment through campaigns on land reform, agroecology and food sovereignty.

Seeds are an essential component of the food system. Indeed, life literally begins with, and remains possible, because of seeds. However, seeds are also a kind of "poster child" for the current industrialized, corporate-driven food system. The increasingly concentrated seed industry, along with the patenting and genetic manipulation of seeds, demonstrates the power of capitalist economic strategies and policies working in lock-step with sophisticated science. This powerful nexus of problems is examined from two different perspectives in section four. Seeds have become one of the most contentious issues in contemporary struggles over food and agricultural production. The challenges of seed sovereignty range from the complex issues of patenting and genetic engineering to the even more complex cultural meanings and customs imbedded in seed exchanges and traditional seed selection and development systems. Access to and control over seeds is an essential building block of food sovereignty.

In the final section, we return to the orienting theme of the book: bridging the gap between the theory and practice of food sovereignty. It is not enough to understand the crisis of the current food system. Although such understanding is a necessary step towards building more sustainable, healthier alternatives, the way forward involves engaging in both resistance and reorientation on many fronts. Thoughtful analysis, a deeper understanding, a commitment to human rights and dignity and an engagement in practical movement politics are all part of the way forward.

The idea of food sovereignty was initially introduced by La Vía Campesina to express both the truth of power relations within the food domain and the hope for the democratic, widely dispersed, just distribution of those powers over food. The term itself opens the way for both critique and hope. This generous, provocative opening has been used to good effect — broadening, deepening, challenging and exploring some key issues evoked by the concept of, and struggles for, food sovereignty. The range and depth of discussions in this book demonstrate both the importance and complexity of food sovereignty.

It is fitting that the idea of food sovereignty was first seeded, metaphorically speaking, by those whose lives and livelihoods are on the frontlines of the battle for control over the land, resources and seeds necessary for food production. The ongoing experience of planting seeds, a long history of struggle for land, resources and social space and the critical, collective analysis of their immediate or imminent displacement have combined to lend peasants and small-scale farmers a particularly urgent perspective on the current food situation. Hence the leadership role that the progressive agrarian movements, gathering as La Vía Campesina, have played in launching and living out the struggle for food sovereignty.

But just as most seeds require appropriate, living soils, water and good weather to flourish, the concept of food sovereignty has "come alive" in a historical moment where many others are recognizing that the current food system is not only part of, but actively perpetuating, destructive environmental, social and political dynamics. This awareness and critique is the living ground in which the struggle for food sovereignty is taking root. The solidarity of allies from all walks of life and many sectors of society around the world provides the needed tilth for food sovereignty initiatives. There is ample evidence that the food sovereignty concept is increasingly firmly rooted. The chapters in this book are illustrative of the vigorous debates it is generating in many quarters, engaging a wide spectrum of people on many levels. The roots are robust and growing, sometimes twisted and unpredictable in their reach, complexity and refinement — but definitely both alive and life-giving. Food sovereignty is a radical, provocative, transformative concept with multiple layers of meaning and application. As with the growing, harvesting and reseeding of grains, achieving food sovereignty is an ongoing, regenerative work in progress. This book is an open invitation to the reader to enter the discourse and engage in this vital work.

Notes

1. From its inception women within La Vía Campesina have engaged in an on-going struggle for gender equality and they have taken some exemplary steps to establish gender parity. The dynamics of this are discussed in depth in Desmarais (2007).
2. The following discussion of Ecuador is based largely on Beauregard (2009). All additional references were cited in Beauregard's report. For a good discussion of this coalition's work on food sovereignty see the documentary *Si a la Soberanía Alimentaria*, available at <dailymotion.com/video/x7arkj_si-a-la-soberania-alimentaria_news>.
3. Even before delivering his partial veto President Correa had introduced a new agricultural law that initially was designed to largely benefit agri-business interests (Denvir 2008, cited in Beauregard 2009).

References

Beauregard, S. 2009. "Food Policy for People: Incorporating Food Sovereignty Principles into State Governance." Senior Comprehensive Report, Urban and Environmental Policy Institute, Occidental College, Los Angeles, April. Available at <departments. oxy.edu/uepi/uep/index.htm>.

Denvir, D. 2008. "Wayward Allies: President Rafael Correa and the Ecuadorian Left." North American Congress on Latin America, July 27. Available at <nacla.org/node/4826>.

De Schutter, O. 2009. "The Right to Food and a Sustainable Global Food System." Special Rapporteur on the Right to Food to the 17th Session of the UN Commission on Sustainable Development.

Desmarais, A.A. 2007. *La Vía Campesina: Globalization and the Power of Peasants.* Halifax, NS: Fernwood Publishing.

Dosh, P., and N. Kligerman. 2009. "Correa vs. Social Movements: Showdown in Ecuador." North American Congress on Latin America. September/October. Available at <nacla.org/node/6124>.

El Universo. 2009a. "Veto Parcial a Ley de Soberanía Alimentaria." March 21. Available at <eluniverso.com>.

_____. 2009b. "Correa Dice que ONG se Meten en Asuntos de Política." March 11. Available at <eluniverso.com>.

FAO (Food and Agriculture Organization of the United Nations). 1999. *The State of Food Insecurity in the World 1999.* Rome.

_____. 2001. *The State of Food Insecurity in the World 2001.* Rome.

_____. 2009. "1.02 Billion People Hungry: One Sixth of Humanity Undernourished — More than Ever Before." FAO press release. Available at FAO Media Centre <fao.org/news/story/en/item/20568/icode/>.

Government of Ecuador. 2008. "Constitutión de la República del Ecuador." Available at <asambleaconstituyente.gov.ec/documentos/constitucion_de_bolsillo.pdf>.

Handy, Jim. 2007. Intervention at the international workshop entitled "Food Sovereignty: Theory, Praxis and Power." University of Saskatchewan, Saskatoon, November 17–18.

International Workshop on the Review of the Agreement on Agriculture. 2003. *Towards Food Sovereignty: Constructing an Alternative to the World Trade Organization's Agreement on Agriculture.* Report of International Workshop, February, Geneva.

La Vía Campesina. 1996a. "Proceedings from the II International Conference of the Vía Campesina." Brussels: NCOS Publications.

_____. 1996b. "The Right to Produce and Access to Land." Position of the Vía Campesina on food sovereignty presented at the World Food Summit, November 13–17, Rome.

_____. 2008a. "An Answer to the Global Food Crisis: Peasants and Small Farmers Can Feed the World!" Jakarta, May 1. Available at <viacampesina.org>.

_____. 2008b. "Declaration of Maputo: V International Conference of La Vía Campesina, Maputo, Mozambique." October 19–22.

Morrison, D. 2008. "B.C. Food Systems Network Working Group on Indigenous Food Sovereignty: Final Activity Report." Prepared for Provincial Health Services Authority — Community Food Action Initiative, Interior Health — Community Food Action Initiative and the B.C. Food Systems Network — Working Group on Indigenous Food Sovereignty, March.

NGO/CSO Forum for Food Sovereignty. 2002. "Food Sovereignty: A Right for All." Political Statement of the NGO/CSO Forum for Food Sovereignty, Rome, June 13. Available at <222.croceviaterra.it/FORUM/DOCUMENTI520DEL%20FORUM/political%20statement.pdf>.

Nicholson, Paul. 2007. Intervention at the international workshop entitled "Food Sovereignty: Theory, Praxis and Power." University of Saskatchewan, Saskatoon, November 17–18.

Nouminren. 2006. "Draft Declaration of Food Sovereignty for the Japanese Farmers and Consumers." Position paper presented at Nyéléni 2007 by the Japanese National Coalition of Workers, Farmers and Consumers for Safe Food and Health, February 23–27, Selingué, Mali.

Nyéléni. 2007. Proceedings of the Forum for Food Sovereignty held Selingué, Mali,

February 23–27.

Peña, K. 2008. "Putting Food First in the Constitution of Ecuador." FoodFirst, October 31. Available at <foodfirst.org/en/node/2301>.

People's Food Sovereignty Network. 2001. "Peoples' Food Sovereignty Statement." Available at <peoplesfoodsovereignty.org>.

United Nations. 2007. *World Population Prospects: The 2006 Revision and World Urbanization Prospects: The 2007 Revision.* Population Division of the Department of Economic and Social Affairs of the United Nations Secretariat. Available at <esa.un.org/unup>.

World Forum on Food Sovereignty. 2001. Final Declaration. Havana, Cuba. September 7. Available at <ukabc.org/havanadeclaration.pdf>.

Ziegler, J. 2004. "Report Submitted by the Special Rapporteur on the Right to Food, Jean Ziegler, in accordance with Commission on Human Rights Resolution 2003/25." United Nations Commission on Human Rights, sixtieth session, February 9.

_____. 2003. "Report by the Special Rapporteur on the Right to Food: Mission to Brazil." United Nations Commission on Human Rights, fifty-ninth session, January 3.

2

Framing Resistance
International Food Regimes
& the Roots of Food Sovereignty

Madeleine Fairbairn

The extraordinary power of language and ideas to animate efforts for social change has long been noted by activists and academics alike. The way that movements frame their ideas influences their likelihood of success as well as the very form taken by their struggle. In the decade and a half since it was first coined, the term "food sovereignty" has already demonstrated great transformative potential as a vision statement and a rallying cry, as a means to stimulate thought and to inspire action. But why did this concept appear when it did? How is its sudden emergence and powerful impact to be understood?

In answering these questions, it is crucial to understand that food sovereignty did not arise in a conceptual or political vacuum. Rather it is one of several extant discourses through which the question of access to food may be understood. In particular, food sovereignty is both a reaction to and an intellectual offspring of the earlier concepts of the "right to food" and "food security." Though all of these concepts are in current use and may therefore be viewed as contemporaries, they were each created at a different point in world history and reflect the particular political and economic climates within which they took shape. In order to fully understand the emergence of the concept of food sovereignty, it is therefore necessary to examine the specific historical conjuncture within which it arose and to locate it in the context of past food-related concepts.

In contextualizing food sovereignty, I make use of two very different theoretical approaches. One is drawn from the writing on international "food regimes," which describes the political and economic structures that undergird successive periods of stability within the world food system. The other stems from the concept of "framing," which examines how social movements deploy language and ideas to mobilize support for their efforts. Harriet Friedmann (2005) suggests a link between these approaches. When a food regime enters into crisis, she argues, social movements explicitly name the previously implicit workings of the regime that have ceased to function properly, thereby deepening the crisis. While Friedmann focuses on how social movements name and rename the *problems* of a food regime in crisis, her analysis implies the possibility of a parallel project, which frames potential *solutions* to the crisis and thereby contributes to regime construction. In other words, framing may also be an *aspirational* project. From this perspective, social movements compete with a host of other actors, including states, corporations and international institutions, over the naming and interpretation of food-related frames that convey their own distinctive visions of how the food regime ought to be structured.

In this chapter I attempt to better understand the recent emergence of food sovereignty by examining how food regimes and food frames have shaped one another throughout the twentieth century. Each of the frames I examine can be seen as an attempt to shape the regime within which it arose. However, forged within the discursive confines of their own particular time periods, they also reflect the structures and ideologies of the very regime they seek to influence. Food sovereignty, too, is a product of a specific historical juncture and food regime. But, as the first frame created by the oppressed rather than the powerful in the world food system, it is unique in attempting to demolish the regime within which it arose and to construct an entirely new one in its place. While all of the food-related concepts examined here aim at addressing the problem of hunger, food sovereignty is the only one to attack it at the root by drawing attention to its systemic causes and striving to create a radically different food system.

The first two discursive frames I examine — the "right to food" and "freedom from hunger" — emerged during the food regime that immediately followed World War II. When this postwar regime entered into crisis, a new frame, "food security" emerged in tandem with the changed political-economic context. In the 1980s the food security frame was reconceptualized as "household food security," which reflects the neoliberal ideology of the emergent corporate food regime. Finally, we see the development of "food sovereignty," which, though itself an organic product of the corporate food regime, reframes the issues so as to explicitly reject its structures and ideologies. Analyzing food sovereignty in this way, as the most recent in a progression of historically embedded frames, allows for a better understanding of both its potential to effect change and the challenges it faces.

Conceptual Framework: Food Regimes & Neoliberalization

The concept of "food regimes" was developed by Harriet Friedmann and Philip McMichael (1989) as a means of linking periods of capitalist accumulation to the international relations of food production and consumption that accompany them. Though they are characterized by relatively stable political and economic relationships, these regimes are nonetheless temporary and historically contingent. The first food regime, encompassing roughly the period from 1870 to 1914, was structured by international free trade between European colonial powers and the settler colonies under British hegemony. Commercial family farmers in the settler states became the primary suppliers of cheap grain and meat to feed the growing urban workforces of European nations. Meanwhile the colonies and former colonies of the Global South supplied Europe with tropical commodities but remained largely self-sufficient in food grains. With the Great Depression of the 1930s and World War II, the regime finally collapsed due to the failure of the gold standard, the demise of free trade and the conflict between the emerging nation-state system and the old colonial system (Friedmann and McMichael 1989).

The crisis lasted for only about thirty years before a new regime had consolidated, this time under the auspices of U.S. rather than British hegemony and based on state intervention rather than free markets. During this postwar food regime, relations of production and consumption were focused on the protection of national markets as nation states followed the example of the U.S. in instituting measures such as domestic price supports for farmers, protective tariffs and export subsidies. Large U.S. agricultural subsidies led to a transformation of markets and diets in the Global South as the federal government began disposing of chronic grain surpluses in the U.S. through the mechanism of food aid. This ingenious arrangement, codified in 1954 as *Public Law 480* (PL 480), served the dual purpose of winning allies in the Global South during the Cold War and disposing of surpluses in such a way as to cause dependency and create future markets for those grains (Friedmann 1993).

The food aid approach to disposing of grain surpluses was particularly effective because the postwar food regime corresponded with what McMichael has termed the "development project" (McMichael 2007). In this state-building process, developing countries saw the particular trajectory of the U.S. and other settler states as universally desirable and universally achievable through the adoption of mercantilist policies similar to those pursued by the U.S. This meant that agriculture was conceived of as a national economic sector characterized by production subsidies, economic regulation and government purchases (McMichael 1992). In the realm of industry, developing nations used import substitution industrialization strategies to protect domestic production. This led to rapid urbanization and dwindling rural populations, which further increased the allure of cheap U.S. wheat and caused many Southern countries to begin importing much of their food supply (McMichael 2007).

The postwar food regime came to an abrupt end in the early 1970s due to a combination of factors. By this time, the food aid program had done its job of creating markets very effectively, and much of the Global South was highly dependent on imported U.S. wheat. As long as the Cold War continued, preventing the U.S. from trading with the massive potential market of the Soviet Bloc, this situation was sustainable. In 1972 and 1973, however, in response to a major shortfall in Soviet wheat production, the Nixon administration sold thirty million metric tons of grain to the Soviet Union, definitively ending the era of grain surpluses and contributing to the world food crisis of 1972–73 (Friedmann 1993).

Another cause of the decline of the postwar food regime was that both agriculture and finance were rapidly outgrowing the social and spatial limits of the nation state as they became international in scope. The simultaneously increasing power of transnational corporations reflected and intensified growing resistance to the protective national economic policies, which had characterized the postwar food regime and the development project in general. McMichael has labelled this new trend the "globalization project," an ongoing process characterized by neoliberal ideology and the transfer of certain powers from nation states to international financial institutions and corporations (McMichael 2007). The welfare state programs, which had been encouraged under U.S. hegemony, are being dismantled as the state takes on a new role in facilitating corporate investment and the liberalization of markets.

The globalization project is the foundation of the emerging "corporate food regime," which aims at the removal of social and political barriers to the free flow of capital in food and agriculture and is institutionalized through international agreements such as the WTO's Agreement on Agriculture. For Southern agricultural production, the corporate food regime means an increasing focus on industrial, export-oriented agriculture, increasing the dependence of farmers on transnational agrifood corporations for their means of production, and the ongoing dispossession and displacement of peasant populations and their cultures of provision (McMichael 2005).

Friedmann (2005) stresses that this new regime involves the selective appropriation by transnational corporations of certain activist demands, leading to a new round of accumulation in the form of "green capitalism." She particularly emphasizes the role of privately enforced quality standards, which corporations use to mollify privileged consumers with new products that promise higher quality or more ethical production. She also suggests that this regime may never fully consolidate, as it is already contested by the consumer activists whose demands it has incorporated, as well as by peasant and small farmer organizations, which pose a very different challenge in the form of food sovereignty.

Theoretical work on neoliberalism is eminently useful in understanding the ongoing shifts in the political economy of food and agriculture that contribute to the formation of the corporate food regime. David Harvey defines neoliberalism as "a theory of political-economic practices that proposes that human wellbeing can best be advanced by liberating individual entrepreneurial freedoms and skills within an institutional framework characterized by strong private property rights, free markets, and free trade" (2005: 2). Many have noted that this policy framework is accompanied by a powerful neoliberal discourse. This discourse has both naturalizing and self-actualizing tendencies; it depicts neoliberalization as an inevitable, external force rather than an intentional project (of capitalist corporations and governments) and it reshapes the social world to fit the picture that it describes (Bourdieu 1998; Peck and Tickell 2002). This has the effect of depoliticizing the process, thereby making the task of resistance far more difficult (Peck and Tickell 2002).

Resistance is also made difficult by the power of neoliberal discourse to influence the everyday language and actions available to people. Patricia Allen and Julie Guthman (2006) argue that even agrifood activism may inadvertently reinforce neoliberal thinking. They suggest, for instance, that the labelling schemes (i.e., organic, locally grown, fair trade) advocated by activists rely on a neoliberal discourse of consumerism, personal responsibility and choice, thereby contributing to the normalization of neoliberalism. This discourse distills resistance down to a matter of individual purchasing decisions, thereby creating a form of food politics which is actually highly apolitical. This reflection reinforces Friedmann's observation that the corporate food regime is particularly difficult to resist because of its tendency to co-opt activist demands.

Frames for the "Free World": The Right to Food & Freedom from Hunger

Although public demands for food have been documented as stretching back many centuries, an internationally acknowledged frame for the expression of the human need for food did not arise until the mid twentieth century. Historians have chronicled bread riots and other public claims of a state obligation to provide food during the first food regime and earlier (Thompson 1971). However, it was not until the consolidation of the postwar food regime in 1947 that the first universally recognized frame for food access emerged in the form of the "right to food." A second, less politically powerful but perhaps more widely used framing of the issues as "freedom from hunger" also emerged during this period. The form taken by these frames was in no way inevitable. Rather, the right to food and freedom from hunger, as frames, were the products of the political-economic context within which they emerged and were shaped by the structures and ideologies inherent in the newly consolidated postwar food regime.

The ideological link between the right to food and the postwar food regime is perhaps most evident in the central political role played by the nation state and the acceptability of state intervention in markets. During the wars, European governments had assumed previously unheard of responsibilities in the regulation of the food supply, undertaking a range of market interventions such as rationing, subsidized bread and nutritional education (Helstosky 2000). This precedent, combined with the acute food shortages experienced by many European countries after World War II, led to the creation of the right to food in paragraph one of Article 25 of the *Universal Declaration of Human Rights* (UDHR) enacted by the United Nations (UN) in 1948. The UDHR provided the basis for the *International Covenant on Economic, Social and Cultural Rights* (CESCR), which imbues the right to food with a legal force (absent from the subsequent frames of food security and food sovereignty). Although the international agreements themselves have no mechanism for enforcement, they create a legal precedent within the ratifying countries, which allows citizens or courts to charge states with the obligation to ensure adequate food in situations in which citizens are unable to secure it for themselves (Alston 1984). These documents were drafted by international bodies, but the obligation to ensure the right to food, as with human rights in general, remains closely tied to citizenship rights and is dependent on enforcement at the national level (Shafir 2004).

The Cold War ideology that structured the postwar regime also played a decisive role in the development (or lack thereof) of the right to food and freedom from hunger. After the creation of the UDHR, a fissure developed between the two different classes of rights it had laid out; while the United States favoured the negative "civil and political rights" (what an individual has the right to be free from), the Soviet Union favoured the positive "economic, social and cultural rights" (what an individual has a right to). The refusal by each side to acknowledge the class of rights espoused by the other meant that, when it came time to elaborate on the meaning of human rights, the United Nations drafted two separate international human rights covenants. The positive right to food falls under Article 1 of the CESCR, which was adopted in 1966 but never ratified by the U.S. Like many other human rights, it would have to wait until the Cold War wound down to become the topic of renewed interest beginning in the 1970s. Ultimately, Cold War politics seriously limited the effectiveness of the right to food as a frame for achieving a more favourable world food situation.

Refusal by the U.S. government to accept any obligation to ensure economic, social and cultural rights is probably the reason that it became common during this time period to refer to "the right to freedom from hunger." By posing the concept in such a way, interested organizations made the positive right to food sound more like the negative rights espoused by the U.S., thereby increasing their chances of attracting international support. A major Food and Agricultural Organization (FAO) campaign, initiated in 1960 and directed at fulfilling the right to food, was tellingly called the Freedom from Hunger Campaign. This campaign inadvertently caused a discrepancy in the CESCR. Whereas the right to adequate food is referred to in Article 11, paragraph one, the right to freedom from hunger is referred to in paragraph two. The FAO proposed the wording of paragraph two in order to lend legal force to its freedom from hunger campaign. Subsequently, however, the two different wordings in the covenant have led to debate over the extent of state obligation implied by the article (Alston 1984). This debate reflects the difficulty that organizations like the FAO faced in framing their goals so as to garner maximum international support within the ideological constraints of the period.

Regardless of whether the frame used was the positively focused "right to food" or the negatively focused "freedom from hunger," the food-related discourse of this period reflects several other dimensions of the postwar food regime and development project. First of all, they embrace the belief in U.S.-style development as a universally attainable goal to be achieved through industrialization and advances in agricultural technology. In a 1968 speech, Addeke Boerma, the director general of the FAO, expressed this belief: "The three basic ingredients required are capital, technology and organization. They are becoming available in increasing measure, and I believe that the agriculture of developing countries is now reaching the point of 'take off'" (Boerma 1976: 21). The reference to "take off" echoes the stages of economic growth conceptualized by modernization theorist Walt Rostow (1960), one of the most influential thinkers of the development project. The approach to development it reflects saw Southern countries as constrained by their "traditional" smallholder agriculture and urged timely injections of capital and agricultural technology as a sure means to launch them down the development path taken by the U.S. and keep them from joining the communist bloc.

Paradoxically, the unique structure of the postwar food regime led to an emphasis on the necessity of food aid alongside the development of national agriculture. The manifesto produced at the Special Assembly on Man's Right to Freedom from Hunger, held by the FAO in 1963, demonstrates this unique combination, stating: "No development can be lasting which is not based on a mobilization of national resources. But external aid is indispensable initially to guide and supplement these efforts" (FAO 1963). Despite the dependency that food aid was creating in developing countries, PL 480 served U.S. interests too well to be left out of the fight for "freedom from hunger," no matter how contradictory its effects.

Whether employing the right to food or framing food issues in terms of freedom from hunger, writings on food from the period clearly reflect the influence of the postwar food regime. Cold War ideologies shaped the political terrain within which hunger advocates were operating, limiting how they could frame their ideas. Food policy recommendations therefore assumed the political centrality of the nation state and the importance of national development goals while simultaneously advocating reliance on U.S. food aid. Though the postwar food regime came to a sudden halt with the world food crisis of 1972–73, the language and concepts that characterized the regime would continue to influence the framing of food-related issues for several years to come.

Regime Crisis & Reconceptualization: The Emergence of Food Security

Friedmann's (2005: 232) observation that "when names catch on, it is a sign that the regime is in crisis" was manifested particularly clearly in the framing of food security at the World Food Conference in 1974. The conference was held in response to the world food crisis that had begun in 1972, when several parts of the world experienced poor harvests, leading to the massive sale of U.S. grain to the Soviet Union, which effectively eliminated the U.S. grain surplus. These conditions were compounded by the failure of the Peruvian anchovy catch, which greatly reduced the supply of the fish meal used in animal feed, and by the 1973 oil embargo, which increased the cost of industrial agricultural production. The world food crisis not only brought an end to the postwar food regime but also forced international leaders to re-evaluate their accustomed approach to food and hunger. In using the term "food security," the delegates to the convention were creating a new frame for food issues that could be used as a tool in forging a new food regime for the future. From this point onwards, food security became the dominant frame for global food issues. Over the coming years, however, shifting political ideologies and food regime rules would shape and reshape the contours of the food security frame, altering how it was both defined and pursued. Though the initial version of the food security frame was already being replaced by the early 1980s, it is worth examining as a reference point for the food security frame as it stands today. The contrast between the earlier and later versions of the same frame demonstrates how sensitive such concepts are to the political and ideological changes going on around them. Though the framers express the aspiration for a better international food regime, they were and are solidly grounded within the reality of existing conditions. As such, the initial version of food security displayed a heavy reliance on the ideas and language characteristic of the postwar food regime whose failure had brought it into existence.

Food security first entered into the official discourse in the report of the World Food Conference of 1974, which called for an International Undertaking on World Food Security. This meeting was hosted by the FAO in Rome and was attended by representatives of 135 countries. Like the right to food and freedom from hunger, food security was conceptualized in the corridors of global power; thus while it attempts to remedy a faulty system, it does so without questioning the dominant political-economic wisdom. Though food security isn't explicitly defined in the report, the term is used extensively. Article G of the Universal Declaration on the Eradication of Hunger and Malnutrition, contained in the report, states:

> The well-being of the peoples of the world largely depends on the adequate production and distribution of food as well as the establishment of a world food security system which would ensure adequate availability of, and reasonable prices for, food at all times, irrespective of periodic fluctuations and vagaries of weather and free of political and economic pressures, and should thus facilitate, amongst other things, the development process of developing countries. (UN 1974)

It is immediately clear that the early discourse surrounding food security was rooted in the ideology of the failing postwar food regime. Rather than prices being determined on a free market, "reasonable prices" must be ensured (presumably by national governments) without hindrance from natural, political or economic circumstances. Ensuring macro-level food availability in this way is justified by its facilitation of national development goals. Throughout the report, food security is discussed in the context of national food stocks and national development. When it is not discussed at the level of the nation state, it is in reference to the maintenance of sufficient food stocks at the international level through cooperation between sovereign national governments.

This state-centric view of food security is also evident in the definitions of food security employed during the 1970s and early 1980s. A 1981 FAO report explained: "Food security in its broadest sense is the availability of adequate food supplies now and in the future. In the narrower sense, food security means food stocks and arrangements to govern their establishment and use as a protection against crop failures or shortfalls in imported food supplies" (1981: 114). In this definition, states are implicitly given responsibility for ensuring a steady food supply. Similarly, a 1977 World Bank staff working paper gives the following operational definition of food insecurity: "the probability of food grain consumption in developing countries falling below a desired level due to a fixed upper limit on the food import bill they can afford and an unfavourable combination of poor harvests and world food grain prices" (Reutlinger 1977: 1). Once again, the emphasis is on national-level supplies, and the major factors involved are national imports and production levels. This emphasis on food availability at the national level went hand in hand with policy proposals that involved strong state action, often including government intervention in markets.

Finally, early food security documents maintain the somewhat contradictory promotion of both industrialization of national agriculture and external food aid. However, they also exhibit an incipient awareness that this arrangement is problematic. The report of the World Food Conference contains a resolution that calls for bolstering the World Food Program, whose primary task was to provide food aid. It states, however, that this aid should be administered in a fashion that is "consonant with the sovereignty rights of nations" and will not hinder national agricultural development or "act as a disincentive to local production" (UN 1974: 15). Though the food aid system was obviously no longer providing a functional basis for the international food regime, it had not yet become clear what would replace it.

To a large extent, the discredited structures and ideologies of the postwar food regime can still be found in the earliest incarnation of food security. Framed when the postwar regime had just barely entered into crisis, it is not surprising that the concept of food security originally drew from the discursive field created by that regime. The concept of food security therefore originally centred on how nations could better control their food supplies through market intervention, increased production and external food aid. However, new structures and ideologies would soon emerge, filling the void left by the regime's collapse and giving food security framers a new set of concepts to work with.

Individual Access, Neoliberal Means:
The Transition to Household Food Security

In the late 1970s and early 1980s, a major shift began to occur in the discourse surrounding food security. The household food security frame that emerged has individual purchasing power at its analytical core and is coupled with policy prescriptions that favour liberalized agricultural markets and a decreased role for national governments. As such, it strongly reflects the neoliberal discourse and structure of the globalization project and nascent corporate food regime (McMichael 2005).

It is impossible to pinpoint the exact origin of this shift as it appears almost simultaneously among academics working on issues of food security and in the international institutions that had originated the concept. To understand the nature of this reconceptualization, it is perhaps best to begin with the work of the Nobel Prize winning economist Amartya Sen. In his groundbreaking 1981 book *Poverty and Famines*, Sen analyzed the causes of several major famines to prove that national-level food *availability* does not necessarily translate into household-level food *access*. Using his own concept of "food entitlement," Sen demonstrated that major famines can occur in countries where overall food availability is sufficient simply because a certain region or occupational group suddenly loses the economic ability to obtain food. If a household cannot afford to buy food (suffers from entitlement failure), its members will go hungry regardless of the overall food availability in the country. Furthermore, Sen pointed out that many households have differential food allocation among their members, with men frequently having far greater access to food than women and children. In other words, even the household might be an insufficiently micro-scale for capturing the effect of differential food access.

The two major theoretical shifts contained in Sen's work — a shrinking of the scale of analysis and a focus on economic access to food — were also underway at the UN and World Bank and are apparent in their notion of household food security. Today, the most basic and commonly used definition of the term is that of the World Bank: "Access by all people at all times to enough food for an active and healthy life" (1986a: 1). In stark contrast to how food security was conceptualized in the immediate aftermath of the 1972–73 world food crisis, this definition hinges on individual access rather than national-level availability. These insights had major repercussions for food security policy as well as measurement, which underwent a parallel shift in focus from national supply to individual caloric intake.

Though it represents an advance in the definition and measurement of food security, this new conceptualization also echoes the neoliberal discourse of the globalization project, which had begun taking shape in the 1980s. Food security is now a frame about the micro-economic choices facing individuals in a free market, rather than the policy choices facing governments. The FAO makes its growing emphasis on individual choice explicit in a 1997 publication: "Food security is as much about individual strategies for survival and wellbeing as about national programmes and public investments in food production and income generation" (1997: 3). This focus on the market decisions of individuals is part and parcel with the hegemonic neoliberal ideology (Peck and Tickell 2002). It also goes hand in hand with an unquestioning treatment of food as a commodity. A 1993 World Bank report, *Overcoming Global Hunger* sums up the implications of the new food security frame nicely:

> In practice, however, food is a commodity. Access to it is largely a function of income and asset distribution, as well as of the functioning (or malfunctioning) of food production and marketing systems. From this perspective, access to food is governed by the same factors that govern access to any other commodity. It is for this reason that hunger and poverty are so closely linked. (1993: 134)

The neoliberal discourse of the globalization project, with its individualizing and commodifying tendencies, had, by the 1990s, ousted the state-centred discourse of the development project as the primary influence on food security.

Perhaps the most striking manifestation of neoliberal logic within the household food security frame, however, is that this need for individual purchasing power is effortlessly transformed into a call for liberal trade policies. It is frequently asserted that only free markets can produce the economic growth needed to increase household incomes sufficiently. Unlike Sen, who advocates a combination of both market and public action to improve household food security, the World Bank was an early convert to this school of thought. Its 1986 *Development Report* repeatedly stresses the negative impacts of government intervention in agricultural markets while devoting relatively little space to the subjects of buffer stocks compared to a few years earlier (1986b). The World Bank continues to promote this neoliberal perspective on food security, which requires that countries "refrain from costly self-sufficiency policies and specialize in producing the commodities which are most profitable for them" (1996: 1). The free trade policies pursued by the WTO also played a prominent role in the neoliberalization of food security. As McMichael (2005: 276) observes, "The shift in the 'site' of food security from the nation-state to the world market was engineered during the Uruguay Round (1986–1994)" of WTO negotiations.

The FAO's use of food security, although less blatantly influenced by neoliberal doctrines, now also emphasizes market orientation over state intervention. National governments, though still prominent actors, are assigned a drastically different role than in the first iteration of food security. Rather than directly shaping markets and controlling food supply,

> governments have a key role to play in creating, through correct policies, an environment which encourages investment leading to food security. This environment is characterized by political stability, good infrastructure, liberal

trade policies, an effective legal framework and social safety nets for the poorest. (FAO 1997: 16)

Though the FAO has not abandoned its traditional stance that national governments should intervene on behalf of their most vulnerable citizens, an important function of states is now to attract foreign investment and enable the smooth functioning of free markets. Similarly, one of the seven commitments contained in the Plan of Action produced by the 1996 World Food Summit is entirely devoted to ensuring that WTO member countries will not renege on their commitments to liberalize trade in their pursuit of food security. Even at the UN, free trade seems to have become an unquestioned priority.

Though it appears to constitute a straightforward reconceptualization, the new framing of food security is remarkably consonant with the neoliberal ideology that emerged at around the same time. The language of household food security demonstrates that the commodifying influence of the corporate food regime, which Friedmann (2005) observes among rich consumers, is also at work among poor consumers. In addition, the transfer of power over food security from the state to global financial institutions like the WTO adheres closely to McMichael's conception of the globalization project, in which "the national coherence of states is being eroded by the internationalization of economic relations, supported by the orthodox liberal ideology of free markets" (1992: 344). Agricultural markets are now entrusted with maintaining food security while the state's primary responsibility is to ensure an economic climate that attracts foreign investment. This considerable and rapid departure from the original conception of food security stems from the changing global political-economic climate and parallel emergence of a new food regime.

Counterframe for the Corporate Food Regime: The Food Sovereignty Movement

During the mid 1990s a new frame arose to challenge the dominance of food security and to envision a more just and sustainable system than that of the corporate food regime. Returning to Friedmann's observation that new names are created when a food regime starts to falter, food sovereignty can be seen as a name given in reaction to the failings of the nascent corporate food regime. Though not yet fully consolidated, the corporate food regime has already shown itself to have flaws that go far beyond the inability to provide food for everyone. These flaws, which take forms such as social injustice, environmental degradation and the loss of traditional knowledge, are repeatedly named and denounced within the food sovereignty frame.

It can be argued that food sovereignty has been just as heavily influenced by the existence of the corporate food regime as was food security. The difference is that, while food security incorporates many aspects of the regime and the globalization project, food sovereignty embodies a self-conscious rejection of almost everything they stand for. It can be seen as a "counterframe" (Benford and Snow 2000) to food security — an alternative schema for understanding the corporate food regime that is conditioned by the very different experience and interests of the framers. Food security, like the right to food and freedom from hunger, was framed by the global political elite. It is little surprise that in their efforts to reduce hunger, they fail to question the political and economic structures within which they rose to power. In stark contrast, food sovereignty was developed by La Vía Campesina, a movement of small producers, peasants and farm workers based in both the North and South. As such, food sovereignty seeks not just to tweak the existing system but to overhaul it entirely. Its advocates, because they are relatively marginalized actors within the food system, call for an entirely new regime, which they variously refer to as a new paradigm, a new model or an "alternative modernity" (Desmarais 2007; Rosset 2003). They reject food security as the discourse of the powerful and propose in its stead an alternative that more faithfully relays the needs of small farmers and conjures the image of an alternate regime in which these needs might better be met.

The most basic way in which the food sovereignty frame challenges the emerging corporate food regime is by calling into question the micro-economic assumptions upon which it is predicated. First of all, advocates refuse to adopt the individualizing language that shapes the household food security frame. Instead they emphasize peasant solidarity and often assert collective rights and ownership over resources. They also argue against the idea that food can be treated as a commodity. The Havana Declaration of the 2001 World Forum on Food Sovereignty states: "We affirm that food is not just another merchandise and that the food system cannot be viewed solely according to market logic" (WFFS 2001: 2). Such descriptions of food as bearing worth beyond its economic value as a commodity occur throughout the food sovereignty discourse and directly challenge the corporate food regime's most basic assumptions. Finally, the food sovereignty frame calls into question the micro-economic framework of the corporate food regime by placing great value on things with little quantifiable economic worth, such as culture, biodiversity and traditional knowledge.

The framing of food sovereignty also seeks to delegitimize the corporate food regime by questioning the increasingly global-level control of the world food system and demanding instead control at smaller scales. Advocates emphasize the need to relocalize both markets and governance, a reaction to the increasing integration of agricultural markets and growing power of global institutions such as the WTO that has occurred under the corporate food regime (see, for instance, La Vía Campesina 1996). This withdrawal from the global scale takes various forms, ranging from affirmations of the right to indigenous self-determination to the reassertion of control by the nation state. Where the corporate food regime removed the state from its role as primary food supplier and gave it instead the role of facilitating free markets in agricultural goods, the state is reinstated into a central position in the food sovereignty frame. State action is demanded in the form of support for affordable food prices, agrarian reform and rural development programs. In shoring up the state, advocates reassert the need for market regulation and condemn the outcomes of international governance of liberalized markets. In rejecting global-level control by corporations and international financial institutions in favour of democratic control at the national or local level, food sovereignty advocates are rejecting one of the most basic foundations of the emerging food regime.

Perhaps the most powerful challenge the food sovereignty frame poses to the corporate food regime, however, is its politicization of food and agriculture. Food sovereignty advocates explicitly name the actors within the system who benefit by maintaining the food regime status quo and from such supposedly neutral policy tools as the WTO Agreement on Agriculture. This stands in stark contrast to the food security frame, which presents the current agrifood system as the natural outcome of spontaneous market forces and often uses technical terminology and scientific rhetoric, which lend the issues an aura of neutrality. By naming the assumptions and politicizing the power structure behind the corporate food regime, food sovereignty advocates poke holes in this naturalizing narrative and threaten the regime's chances of successful consolidation.

The Potential of Food Sovereignty

The concept of food sovereignty is the most recent in a series of frames created to address global food issues. Each successive frame has been historically contingent, rooted in the existing food regime and influenced by the dominant political and economic ideology (Table 2.1). The right to food and freedom from hunger frames were created during the postwar food regime by relatively powerful diplomats. They therefore incorporated several dimensions of the regime, including the political centrality of the state, the subordination of markets to social concerns, the system of international food aid and the polarizing politics of the Cold War. When crisis struck the postwar regime, delegates to the 1974 World Food Conference reframed the issues in terms of food security. Though this new frame was created in response to the failings of the postwar regime, its language and associated policies continued for several years to reflect aspects of that regime and the development project.

During the 1980s, the food security frame underwent a major reformulation that led to a focus on household incomes and individual food acquisition. Not surprisingly, this shift occurred just as the lineaments of a new, corporate food regime began to emerge from the unstable lacuna left by the breakdown of the postwar regime. Once again the framers were largely in the seat of power, and the household food security frame that resulted therefore reflects several features of the new regime and ascendant globalization project. These features include an increase in the power of transnational corporations and international financial institutions, the further transnationalization of agriculture and the prioritization of market

Table 2.1

Food Regime	Dimensions	Frames
First food regime (1870–1914)	Colonialism Free markets British hegemony Grain trade between settler states and Europe Agricultural industrialization	No universal frames
Postwar food regime (1947–1973)	System of independent nation-states Market intervention/regulation U.S. hegemony Development project Cold War Surplus/food aid complex Agriculture is a national sector Agricultural industrialization	Right to food Freedom from hunger Food security (emerges only with the postwar regime crisis)
Corporate food regime (began emerging in 1980s)	Nation-state loses political centrality Free markets Globalization project Neoliberal discourse and ideology Agriculture becomes an international sector Increasing power of agrifood corporations Incorporation of some elite consumer demands. Agricultural industrialization.	Household food security Food sovereignty (working to create a crisis in the corporate regime and foster a new regime)
New regime (?)	Sovereignty of peoples, communities and nations Democratic control of the food system Priority given to local agricultural markets Trade subordinated to social goals Food and seeds valued as more than commodities Sustainable agricultural production	

liberalization over social goals. The household food security frame continues to dominate the language of international institutions, national governments and most NGOs.

Over the last decade, however, food sovereignty has arisen as a counterframe based on a rejection of most aspects of the corporate food regime. It is a product of the regime in that it could only have emerged within this specific historical, political and ideological context. Framed by the traditional underdogs of the world food system, however, it is the first frame that seeks to overturn the regime within which it was created. Where household food security uncritically assimilates many dimensions of the corporate food regime, food sovereignty categorically rejects them. It emphasizes solidarity over individualism and insists that food cannot be treated as a typical commodity, in contrast to the neoclassical economic assumptions of the food security frame. It criticizes free markets and demands state intervention, in contrast to the calls for trade liberalization that characterize the food security frame. It is also composed of highly political language, which contrasts with the neutral and technical language frequently used in food security documents.

Food sovereignty is unique among its predecessors inasmuch as its proponents actively and self-consciously seek to supplant the regime within which it originated and is still embedded. The right to food (as originally deployed — it is now frequently used by food sovereignty advocates and others in a far more conscientious manner) assumed the necessity of food aid, and household food security fails to question the importance of economic liberalization. By incorporating, intact, the major tenets of the regimes within which they arose, they may have the contradictory effect of displacing peasant producers and aggravating global hunger — the very problem they purport to address. Food sovereignty advocates, on the other hand, attack hunger by challenging the dominant food system itself. In so doing, they demonstrate that the existing regime has not yet and may never reach the hegemonic status of the two regimes before it.

Because it represents a genuine alternative to (rather than just a variant of) the existing model, food sovereignty may be able to withstand the risk of co-optation or dilution. The movement is unique among current agrifood activism in the extent to which it casts off the language of neoliberalism and creates a viable discursive alternative. Allen and Guthman's claim that activism may inadvertently reinforce, rather than hinder "neoliberal subject formation" could be levied, for instance, at all campaigns that use consumer purchasing power as their primary tool. Product boycotts and certification schemes, which use the market as their main mechanism, implicitly legitimize the increased power of corporations in the current food system. Food sovereignty advocates, rather than participating in this type of "green capitalism" (Friedmann 2005), target political bodies, not corporations. The change they demand will be effected through political action by democratically elected bodies at all levels of government, not through corporate codes of conduct or new "ethical" product lines. The intensely political language used by food sovereignty advocates makes it very difficult for their demands to be assimilated by corporations and therefore increases the strength of their challenge to the status quo.

In musing over the impact of anti-globalization protests, Peck and Tickell (2002: 400) remark: "In its own explicit politicization, then, the resistance movement may have the capacity to hold a mirror to the process of (ostensibly apolitical) neoliberalization, revealing its real character, scope, and consequences." Here they point to what is perhaps the greatest potential of the food sovereignty frame. Whereas the right to food and food security frames address the problems of the global food system in legal and economic terms respectively, food sovereignty is framed in highly political language. This allows it to bring to light the power relations that have led to the formation of the food regime and gives it greater transformative potential than its predecessors.

Though food sovereignty is unlikely to encounter any immediate success in instituting the new food regime that its advocates desire, it does have the potential to destabilize the corporate food regime. By explicitly rejecting the assumptions of the current regime and politicizing its ostensibly neutral workings, food sovereignty advocates have already helped to underscore its flaws, thus fostering an awareness that could prevent this regime from successfully consolidating. Situating food sovereignty within the context of the other food-related frames has demonstrated that it is solidly grounded in the political economy of the current world food system but that it nonetheless has the drive and potential to radically alter that system.

References

Allen, P., and J. Guthman. 2006. "From 'Old School' to 'Farm-to-School': Neoliberalization from the Ground Up." *Agriculture and Human Values* 23, 4.

Alston, P. 1984. "International Law and the Human Right to Food." In P. Alston and K. Tomaševski (eds.), *The Right to Food*. Leiden, The Netherlands: Martinus Nijhoff Publishers.

Benford, R., and D. Snow. 2000. "Framing Processes and Social Movements: An Overview and Assessment." *Annual Review of Sociology* 26.

Boerma, A. 1976. *A Right to Food: A Selection from Speeches by Addeke H. Boerma, Director-General of FAO 1968–1975*. Rome: Food and Agriculture Organization.

Bourdieu, P. 1998. *Acts of Resistance*. Cambridge, UK: Polity Press.

Desmarais, A.A. 2007. *La Vía Campesina: Globalization and the Power of Peasants*. Halifax, NS: Fernwood Publishing.

FAO (Food and Agriculture Organization of the United Nations). 1963. *Man's Right to Freedom from Hunger*. Rome.

_____. 1981. *Agriculture: Toward 2000*. Rome.

_____. 1997. *Investing in Food Security*. Rome.

Friedmann, H. 1993. "The Political Economy of Food: A Global Crisis." *New Left Review* 197.

_____. 2005. "From Colonialism to Green Capitalism: Social Movements and Emergence of Food Regimes." In F. Buttel and P. McMichael (eds.), *Research in Rural Sociology and Development*. Oxford, UK: Elsevier Press.

Friedmann, H., and P. McMichael. 1989. "Agriculture and the State System: The Rise and

Decline of National Agricultures, 1870 to the Present." *Sociologia Ruralis* 29, 2.

Harvey, D. 2005. *A Brief History of Neoliberalism*. New York, NY: Oxford University Press.

Helstosky, C. 2000. "The State, Health, and Nutrition." In K. Kiple and K. Ornelas (eds.), *The Cambridge World History of Food*. Cambridge, UK: Cambridge University Press.

La Vía Campesina. 1996. "The Right to Produce and Access to Land." Position of La Vía Campesina on Food Sovereignty presented at the World Food Summit, November 13–17, Rome.

McMichael, P. 1992. "Tensions between National and International Control of the World Food Order: Contours of a New Food Regime." *Sociological Perspectives* 35, 2.

_____. 2005. "Global Development and the Corporate Food Regime." In F. Buttel and P. McMichael (eds.), *Research in Rural Sociology and Development*. Oxford, UK: Elsevier Press.

_____. 2007. *Development and Social Change: A Global Perspective*. Los Angeles: Pine Fore Press.

Peck, J., and A. Tickell. 2002. "Neoliberalizing Space." *Antipode* 34, 3.

Reutlinger, S. 1977. "Food Insecurity: Magnitude and Remedies." World Bank Staff Working Paper no. 267. Washington, DC: The World Bank.

Rosset, P. 2003. "Food Sovereignty: Global Rallying Cry of Farmer Movements." *Food First Backgrounder* 9, 4.

Rostow, W. 1960. *The Stages of Economic Growth: A Non-Communist Manifesto*. Cambridge, UK: Cambridge University Press.

Sen, A. 1981. *Poverty and Famines: An Essay on Entitlement and Deprivation*. Oxford, UK: Oxford University Press.

Shafir, G. 2004. "Citizenship and Human Rights in an Era of Globalization." In A. Brysk and G. Shafir (eds.), *People Out of Place: Globalization, Human Rights, and the Citizenship Gap*. New York: Routledge.

Thompson, E.P. 1971. "The Moral Economy of the English Crowd in the Eighteenth Century." *Past and Present* 50.

United Nations. 1974. *Report of the World Food Conference*. Geneva.

WFFS (World Forum on Food Sovereignty). 2001. "Final Declaration of the World Forum on Food Sovereignty. Havana, Cuba." Available at <fao.org/righttofood/kc/downloads/vl/docs/AH290.pdf>.

World Bank. 1986a. *Poverty and Hunger: Issues and Options for Food Security in Developing Countries*. Washington, DC.

_____. 1986b. *World Development Report 1986*. Washington, DC.

_____. 1996. *Food Security for the World*. Washington, DC.

_____. 1993. *Overcoming Global Hunger*. Washington, DC.

3

Seeing Like a Peasant
Voices from La Vía Campesina

Itelvina Masioli & Paul Nicholson

The concept of food sovereignty emerged out of the La Vía Campesina peasant movement and continues to be developed, enhanced and refined in the cauldrons of practical political and social struggles. Far from merely serving as a medium of nutrition and a marketable good, food is also the locus of highly charged political struggles. The language of food sovereignty is consciously political language. It speaks to the power of decision-making and control over key facets of the food domain, including the basis of and models for production, access to and control over natural resources, ownership and use of knowledge, seeds and the organization of markets.

The issues raised by food sovereignty are many, varied and inter-related in complex ways. Just as biodiversity is necessary for a healthy ecosystem, so a great diversity of philosophical and strategic thinking and debate is necessary to enhance and enliven food sovereignty. As well, the practical and political expressions of food sovereignty are, and will continue to be, diverse because they emerge out of, and are integrated into, a wide variation of local conditions.

In the following conversations, two farm leaders describe their own unique contexts and articulate some of their strategies and struggles for achieving food sovereignty. Their experience, coupled with the analysis and insights they have garnered from those experiences, provide a small window into the on-the-ground struggle for food sovereignty in two vastly different regions.

The interviews with Itelvina Masioli and Paul Nicholson, two leaders of La Vía Campesina, were conducted in different times and places. Hannah Wittman spoke with Itelvina Masioli, member of the National Coordinating Council of the Brazilian Landless Rural Workers Movement (MST) and representative to the International Coordinating Commission (ICC) of La Vía Campesina, in November 2008, in Saskatchewan, Canada. Nettie Wiebe engaged Paul Nicholson, a Basque peasant movement leader, member of the European coordination of farm organizations and a founding leader of La Vía Campesina, in a conversation about food sovereignty during La Vía Campesina's ICC and staff meetings at Nyéléni, Mali, in June 2009.

Itelvina's and Paul's responses to our questions have been amalgamated for ease of reading. Their combined responses serve to highlight both the similarities and diversity of strategies, tactics and political approaches to implementing food sovereignty in different agroecological, cultural and political contexts.

What does food sovereignty mean to your organization?

Itelvina: Food sovereignty for the Landless Movement, I can say also for the group of movements of La Vía Campesina, is the right of peoples to decide and produce their own food. It is a political right to organize ourselves, to decide what to plant, to have control of seeds. Food sovereignty is a very broad concept that includes the right of access to seeds, the right to produce, to trade, to consume one's own foods. Finally, it is a concept that is linked to the autonomy and sovereignty of peoples.

The peoples' struggle for food sovereignty is tied together with the struggle for autonomy. The autonomy of the people then, is within this much broader concept of sovereignty — a sovereignty of having the right to produce culture and all that is linked to truly human autonomy.

In practice, autonomy is a right of communities, of the people to construct their own destiny, with respect to their rights to culture, to their territory, to the goods of nature. This autonomy seeks to direct the goods of your community towards human, social and economic development so that they are at the service of the wellbeing of the majority.

It is impossible to speak of the autonomy of communities, the autonomy of people, in a capitalist patriarchal system, where the neoliberal model has profit at its centre, where everything is transformed into a commodity, including human beings. With autonomy we are speaking of a real possibility of constructing another kind of society that has, in the first instance, other values and other principles based in environmental, social, and economic sustainability. These other principles and values govern economic, social, cultural and political life in this new society.

How does food sovereignty challenge the dominant mode of production in your country and in your region?

Paul: Over the last twenty-five years, we have been under a government which is extremely neoliberal. We come from a small and very mountainous country, making it difficult for us to compete with the major food-producing areas. We are small farmers — forty-hectare, thirty-hectare, twenty-hectare, ten-hectare farmers. So, for us, the food chain is very important. The control by the major distributors of the food chain means that we become "non-competitive" in the neoliberal sense.

For us, food sovereignty means defending local markets and local food production systems. Therefore, for example, we are defending, with all our capacity, the maintenance of local abattoirs, the maintaining of all of our local markets. A very important dimension for us is the relationship between urban and rural society, the relationship between consumers and producers. Consequently, we are also playing a big part in the struggle against the supermarket chains.

For us, food sovereignty is quite closely felt because it means the right of having food and agricultural policies evolving from our own political context and for our communities. A key issue for us is land use. The question is: what is the land for? There is a major contest between the industrialization and urbanization of land use versus its food producing use. So, we use food sovereignty as the umbrella concept to defend the right of consumers to consume local food as opposed to the imposition of global food. We use food sovereignty to defend our right to maintain a family farm food-producing system, which is sustainable and manageable.

Itelvina: In Brazil and Latin America we are speaking of a class struggle, of the dominant class and the working class, because it is linked exactly to two models, or better said, to two social projects. In the majority of Brazil and Latin America, what is in place is a model tied to agri-business and the transnationals, and every-thing that signifies the implementation of the neoliberal model. This is a model that kills life in a general sense.

For us, food sovereignty is a class struggle that confronts this model of so-ciety, and it comes from the community, from localities, in the organization of community, in efforts to construct another social and political consciousness of participation.

In practice, it is participation; it is ourselves included, because things do not take place in an already established form. How the community begins to organize, from having a culture that is representative democracy to a participatory democ-racy, where the social and human subjects who live there are part of a community that constructs an identity and that then, in its life, in its form of producing, in its cultural life already starts to produce other cultures and values. This entails recuperating one's seeds — it is in fact to feed oneself with what is produced. It is to go on constructing other mentalities of consumption. It is an arduous struggle, because the current system destroys human beings and places its form of life, its form of thinking in the heads of the working and peasant class.

Because of this, we believe that it is an ongoing struggle of constructing an-other mentality in practice. This other mindset is married to another form of social organization that goes beyond simply making of a difference now to engaging in a long-term struggle of education of the new generations.

For example, it is one thing for us to hold a theoretical viewpoint. But we are being bombarded — our children, our families, ourselves — we are bombarded twenty-four hours a day by a cultural industry. So we are constantly challenging ourselves to constructing our theory through practice. And this means that if you do not have a political, economic and social consciousness, you speak one thing and practice another, because in practice, we have to start to disown the capitalist counter-values. That is not easy. It is an arduous struggle to ensure our concept of food sovereignty, of the autonomy of peoples, continues to have real power. It can't stay fixed in concepts only.

I think that it is necessary in our everyday lives, from our community and family lives, to put into practice what we believe. The construction of sovereignty goes from the simple, from the small — from practices, attitudes, behaviour, from what I eat, as I plant, as I cultivate in my community — to the major things of the general political struggle.

What are the most significant challenges or difficulties in your region in efforts to implement food sovereignty?

Itelvina: The difficulties are linked to the model of society and how it is structured. Land today is concentrated in the hands of the big companies, the many multinationals and big plantations that continue imposing their model. So our implementation is difficult because the communities are in a permanent struggle of for access to the land — a struggle against this model, against the transnationals.

Today in Brazil, it is a direct struggle against agri-business. And there is no real agrarian reform policy in Brazil, as in the majority of countries in Latin America. What exists most of the time is the distribution of lands in the face of social pressure and social struggle. Then, when the communities, the settlements, the camps, after years of struggle and confrontation, do achieve access to land, the huge majority hardly have enough labour power. When you arrive at the land, it is also difficult. You find a land totally destroyed by capital. In order to make it produce again, it needs investment. It takes time for you to recuperate this land, and it also demands investment.

And with what means are you going to make the land produce? It is a huge struggle then because, upon winning the land, we engage in a whole struggle to make it produce. And to make it produce within another conceptualization, that is, to engage in agroecological production for the recuperation of this land.

Then, there are the difficulties that you continue to confront: the lack of public policies oriented towards peasant agriculture or family agriculture that would provide support so that communities can produce and build food sovereignty. Today, in Brazil, as I see it, all state and federal government policies are oriented towards agri-business. There are a group of measures and laws that are barriers to small-scale and peasant agriculture. For those of us who have the challenge of producing food, it is difficult, because the cost of production is very high. An agricultural policy oriented towards the production of food does not exist. The governments of our countries are committed to the model of agri-business that produces for export. It produces grains now for agrofuels rather than food.

For those of us who engage in the staunch struggle, we say that agriculture and agrarian reform are necessary for the production of healthy foods, that a country cannot be sovereign if it does not produce its own foods, the food crisis comes on top of all the other crises. There are many obstructions that go on making our day-to-day lives and our implementation of food sovereignty difficult.

For us, what is important is to work at the community level, which from its locality has to produce its own foods. It has to confront monoculture, to produce in a differentiated way, to produce all types of foods, to build a local market, to look for direct sales alternatives, as well as exchanges with other communities. These processes take a long time because they depend on organization. This involves local organization, consciousness-raising and the search for a strategy that ties the countryside with the city in an ongoing strategic alliance. It ties those who produce with those who consume. This is a large process in which we already have many signs of advancing. But it also has many barriers that render much larger steps difficult.

Paul: Land use, as I had mentioned earlier, is a major issue. As we are a small country, the available good agricultural land is under pressure. And there we are losing our battle against industrial uses of land — industry, infrastructures, high-speed trains, motorways and second homes. They're taking over the better agricultural land and leaving us the slopes and the very mountainous areas. So, for us, the use of the common goods, especially land, is a very important fight. Certainly land and seeds are key.

We also need political instruments. We need policies which defend family farm based agriculture rather than export-orientated agriculture. And for this we need specific policies, which promote a sustainable or agroecological mode of production.

Our context over the last twenty-five years is basically that what has been promoted is wine-producing export-orientated production and milk-intensive production. So we have been defending, for example, our locally based food production which is of excellent quality. For example, on-farm cheese-making, on-farm wine-making and on-farm food transformation. I, myself, am part of a small cooperative of fourteen farmers who produce individually, but we transform or process that produce together for local markets. We produce, for example, cider, marmalade and jams for a market area of about thirty-five miles. So that is the kind of model we're trying to defend. And that, for us, is food sovereignty.

In addition to your own organization, who else is working towards food sovereignty in your region, and are some of these organizations more successful than others? If so, why?

Paul: The initiative for food sovereignty in the Basque country also has been developed by social organizations like Emmaus. Emmaus is a European movement, which supports the homeless, the structurally poor people.[1]

There is also a big body of non-governmental organizations who defend food sovereignty in the work of development. And we have the environmental movements too. The issue of genetically modified organisms has been helpful in facilitating alliance-building.

And we are working quite well with other farmers' organizations. We are now developing community-supported agriculture together. And, yes, I think, on a social base, we are increasing our influence. And it's certainly not just a farmers' issue; it's a consumers' issue, an environmental issue and a social issue too.

Itelvina: Food sovereignty is an extremely important issue that we have succeeded in working on together, directly with the group of Brazilian social movements associated with La Vía Campesina: the Movement of those Displaced by Dams, the Small Farmers' Movement and the Peasant Women's Movement. We have also worked together a lot here in Brazil on food sovereignty with the World Women's March, and we have spoken of it in the network of social movements and in the union movement. We understand that food sovereignty interests all of society, like the issue of agrarian reform, like the issue of energy sovereignty and like the issue of water. They are all concerns that society together has to appropriate in order to debate and put on pressure. All of these need to be added into society's daily struggle. I can say, without a doubt, this group of movements has integrated food sovereignty into its internal dynamics, into the struggle, into its political education and into daily practice.

We have talked a lot about food sovereignty in the student movement, principally in the areas of agrarian studies. We have also debated it inside of the Unemployed Movement and some states have made a direct association between the purchase of products of agrarian reform settlements directly for the communities of the Movement of Unemployed Workers. This has happened in the South of Brazil and to some extent here in São Paulo, where there are promising initiatives.

The most organized part of the homeless movement is also having this debate, and we have tried to develop a relationship with them. Here in São Paulo, especially, there are very important experiences that come from the land communes. These are settlements in the urban periphery of the big cities; they are small parcels of land where the people have a place to live. This too is work oriented towards food sovereignty because they produce what is eaten. And it is another form of organizing. Producing food on the land is the fundamental step of the construction of autonomy.

So I think that there are many different experiences in other movements that are linked to La Vía Campesina. These experiences, from the practice and debating of the issue of food sovereignty to engaging in actions, contribute to making a difference and go on to generate another outlook for life.

What are the most significant activities and strategies that your organization is engaging in? And which are most successful? Have you had failures, and if so, what led to these?

Itelvina: In the analysis that we have done, an analysis of our successes and failures is very important. The food sovereignty struggle obviously is part of the necessary struggle for general agrarian reform.

We, as a whole group of social movements involved in agrarian reform, have

worked on the concept of cooperation and on how to develop strategies of cooperation. Throughout these twenty-five years, we have developed a strategy of forms
of cooperation that has to be strictly tied to the organization of the community.
We have some experience of constructing big cooperatives in the sense of having
a very big structure. And this generated some difficult moments of development
and of maintaining them in a good condition. Some have stayed secure, and those
have been a reference for us here in the movement. But it appears to me that our
experience has shown that cooperation is related to scale.

We have learned not to preoccupy ourselves so much with constructing very
big structures but more with building more productive structures of small agro-
industry. It seems to me that they are stronger. The small agro-industries, the collective production groups, seem to find different ways of cooperating, of producing
better and of marrying production with social and political organization.

We have to continue to be vigilant. From this point of view, we have to always
be asking: for whom do we want to produce? What do we want to produce? And
how are we going to produce? Because if not, there is the risk that the community
develops production to improve its financial agenda by accumulating profit rather
than engaging in the class struggle of confrontation with the capitalist model. These
are some of our reflections.

It's clearly a principle of our purpose as human beings to accumulate a quality
of life with consciousness. At times, we do not accumulate much because of the
confrontation of our project of society with a huge structure that up until now
produces much more but does not accumulate in terms of consciousness, of food
production, of application of socialist values of solidarity within the community,
of comradeship. Because of this, we always have to be watchful of our proposal of
cooperation. It has to be linked to our strategy to be accumulating in quality, in
consciousness, in the dreams of our project of a new society that we go on strengthening — of the new man and the new woman.

Paul: We have defended the abattoirs well. That has been a success. And it's been
a success because we worked with the local town councils with the argument that
meat is a basic food in our diets and that we have to control it. That it is a public
service, and it is the responsibility of towns and cities and farmers and butchers and
consumers to work together to maintain locally based abattoirs and not depend on
imported meat, over which there is no health control or no social control either.
And that is a victory.

Our defeat has been mostly in the milk sector, where the intensive milk production policies have convinced the farmers through subsidies that that is the way
forward. And obviously now, in a model of high-cost, high-price situation, when
there's a high price, they were happy. But we're now going into a low-price for the
farmers, and in that situation, it's ruinous. There's no way of producing milk with
high cost at low prices.

Another victory that makes us happy has been our organization at the national level, the Coordinadora de Organizaciones de Agricultores y Ganaderos (Farmers and Ranchers Coordinating Council), which has a team of economists producing a monthly report on the costs, the prices farmers are getting at the farm-gate and the prices consumers are paying in the grocery stores. And we're doing follow-up on forty food prices: milk, beef, chicken, pork and all the vegetables and fruits. The differences are between 100 and 400 percent between what the farmers get and what the consumer pays — and the tendency is for these margins to increase.

Through this analysis and data we are developing strong alliances with mainstream consumer organizations. And we are questioning the deregulation and the loss of control by the governments and the public systems on the food chain. For us, this is another very positive element of moving ahead on food sovereignty.

But, finally, the bad news is that we are losing a lot of farmers day-to-day. And not only small family farmers, but big farmers also cannot compete. And in fact, very often, because of the debt levels, they are closing down faster than the small farmers. But we are losing in the Basque country about one and a half farmers every day, and for us, this is very serious. In the European Union, it's one farmer every minute. And this is happening in the face of policies which support agro-industry, the big industrial farmers, the consolidation of food production and the lowering of prices at farm gate.

What kinds of conditions have to exist for the successful implementation of food sovereignty? And how are governments — local and national — responding to the call for food sovereignty?

Paul: In Europe, governments are very liberal. There is no difference between the social democrat parties and the parties of the right or of the conservatives. They are both very, very liberal. What does that mean? Their priority is to open markets, to bring in global food, to negotiate with the corporations for implementing a control system which is privatized and not public. In the food crisis, for example, the governments didn't have any political mechanisms for recognizing the crisis, of controlling the crisis, or reacting to the crisis. All the instruments were privatized.

So, for us, it's very difficult to speak with the national governments, and this is the context of the whole of Europe. Perhaps there are governments which are more sympathetic or where the rural culture politically is more important. For example, Austria is a GMO-free country, where a locally based agriculture is a priority. But even there too they have policies, which say, "we support agro-industry and we support family farm systems." France is another country where they might maintain policies which defend a more sustainable agriculture. But at the same time, they support a decidedly competitive export-orientated agriculture. So they have a dual system policy.

At local levels there might be some better initiatives, but obviously, as I had mentioned earlier, one issue in Europe which is very important is land use. And there the tendency is to liberalize land use: to make land use very market-orientated with very little public conditioning. And that means that land is going into the hands of the rich — the urban rich, who need their second homes, or the savings bank. They use it also as a speculative measure.

I think in Europe it's a very free-trade orientated policy. We won't speak about the United Kingdom, where it is publicly said that farmers are not of interest to the country. But in other parts of Europe, where there is about 5, 7 or 8 percent of the population farming still, the capacity for resistance is easier or better perhaps.

In our region it is difficult for farmers' organizations alone to generate a debate on agricultural policy. Consumers are more concerned about quality and safety. Our main objective then is to generate a preoccupation with policy on food and agriculture among all citizens. There is a main effort now, among our organizations, to engage in a strategy of building alliances. This reflection comes from the Nyéléni process. In Europe now, in the Basque country, we're identifying the diversity of resistances, the practices as well as the demands. In alliance building it is important to take part in dialogue, there's room for internal differences, there are many issues yet to be discussed.

At the Fifth Conference of La Vía Campesina in Mozambique, La Vía Campesina initiated a campaign to end violence against women. What does food sovereignty mean for gender relations?

Itelvina: We have talked and reflected on this. In the movement we can't talk about agrarian reform or food if in fact we do not confront all the patriarchal values that are so strong in our rural societies. Our form of social organization has to also have this view towards the participation of women. We have debated and advanced this: what other kind of society are we going to construct if we do not also break down the barriers that are inside our heads?

We have to see that in the countryside there are men and women, there are adults, the elderly and there are children. Then, we need to strive to develop in ways that the differences that we have as far as sex are not able to continue being inequalities that impede women and men from being human beings.

I think that the issue of food sovereignty is tied directly to, and we confront it in, systems of machismo, of violence, of all the patriarchal values. The social movement, the community, has to continue constructing instruments that go on inhibiting all these patriarchal tendencies, while continuing to construct other principles and values from this perspective of new social relations of gender.

Paul: Certainly women are among the key advocates or pursuers of food sovereignty. That is very clear. But whether, in general, it is thought that food sovereignty is a way of strengthening gender relations, that is another question. I don't think we're really aware of that.

I don't think we're aware of that, because in Europe, one of the strange things is that, though there is certainly gender inequality and violence against women just as there is in all the world, and especially in the farming world, I don't think that it is understood. I think those issues are still taboo in Europe. And certainly it is an objective of putting it on the table in a frank way. But there is a general lack of acceptance by society, even those countries which are supposed to be extremely progressive in this sense like Sweden or Norway, where there is the same amount of gender violence. So I would say that it is not recognized yet. That it's a taboo issue. And Europe thinks those dreadful things are happening in the rest of the world but not at home. And that is certainly a big challenge for us.

La Vía Campesina also has a commitment to support sustainable land use and agroecology. What does food sovereignty mean for the environment?

Paul: I think that is one of the key issues and one of the key areas of food sovereignty where we're capable of integrating large parts of society. The struggle against GMOs is a part of the food sovereignty struggle. And in Europe, in the face of massive propaganda in favour of GMOs, we have managed to organize a resistance which still maintains 80 percent consumer opposition. Not only to GM food consumption, but to GM food cultivation also. For example, this year in France, by law, not one hectare of GM maize is going to be planted. And the only country which still produces GM maize is Spain. And right now there is a split in the European Union governments, whereby the governments of the North of Europe, where they cannot produce GM maize, are in favour of GM maize and those countries in the South, which can produce GM maize, are basically against. So, it is completely contradictory.

And right now there is a very strong mobilization in Spain to ensure that GMOs are not planted for health reasons and for environmental reasons. And the Austrian government, this last week, has presented a case for social and economic reasons. And probably it will have to be accepted by a court in the European Union.

So our perspective for GMO resistance in the short term is positive. And no doubt, we, La Vía Campesina organizations, go closely in hand, for example, with Greenpeace and Friends of the Earth. So it's an important alliance-building area.

But the whole issue of genetic contamination [of non-GM crops and other plant species] is a major issue too. And we're slowly winning that battle too, because it's been recognized that there is contamination and that there is no possible insurance against it. That is one of the main arguments that we're using. So, yes, the link between food sovereignty and environment is important, and GMOs are a perfect example of this.

Itelvina: We have contradictions in our communities. For us, here in Brazil, this is an ongoing struggle. We live in a crisis. Because of the propaganda bombardment, some community or a peasant thinks that it can resolve its crisis alone. "Oh, I am going to rent out a small piece of land in order to plant soya." Now you can't condemn the organization for this. The organization has to do the work in the community so that the community understands what planting the seed means, what it means to plant conventionally, what it means to maybe clear the forest for something else, what the construction of another agroecology, recuperating seeds or of not being dependent on the transnationals means. This is permanent, long-term work at the base.

Without this work, we will not have the sufficient strength to confront the "monster." We are speaking of a struggle that is hard against powerful enemies, like Monsanto and Cargill. Because of this, our struggle, our victory, our advance, is in our capacity as organized people. But it is not just the sum of families, of men and women, but of consciousness. There is no advance in simply being a mass of people together. It has to be people with consciousness. And this is political work of ongoing education.

An enormous diversity of global organizations and movements is working on the concept of food sovereignty in various ways. We have 149 organizations of La Vía Campesina — and there are more organizations that are not linked to La Vía Campesina. What are the challenges in working together, considering distinct cultures and regional agricultural practices? How do you work together considering this diversity?

Paul: Alliance building is complex. It is complex because we're new to the game. I think alliance building is a new political culture, which has developed over the last ten years especially. Before this we were very sectorialized, and each one was on [their] own field and not bothering with anybody else. But right now I think the current is the other way. It's far more horizontal. Political party control is far less. Now it is far more heterogeneous.

And we're learning very fast in Europe to work in alliances, in spaces with like-minded organizations — and with not like-minded organizations too. So in a very contradictory situation, we are building alliances and learning how to work together.

Food sovereignty is going to be one of the main issues around which to build alliances. And here we have to be very clear too. We cannot have alliances which just defend food sovereignty as in: "eat local, eat good." We're speaking of changing policies, of changing economic paradigms.

We are going to define our alliances carefully and to define them in the sense that they must be alliances for change. Not just alliances to go over minor issues or with a shorter-term perspective. We are speaking of building alliances to change these policies and to change society itself.

Itelvina: This is very huge work. It is arduous. And it seems to me that we have advanced in these past few years as La Vía Campesina, especially in the political debate of the concept. We are gaining a greater unity in our understanding of sovereignty — that the right to produce is part of the right to decide. Our concept of sovereignty has to be linked to a project of agriculture. Our point of view has some unity in the concept and unity in its elaboration. We have a huge task in that we are guardians of the land, of the seeds, of the water, of all resources, of all natural resources, of the goods of nature. This also demands a political coherence in our practice.

There have been enormous advances because these projects are not isolated. For example, in the same way as it is very difficult to succeed in implementing a complete agrarian reform within capitalism, we believe food sovereignty can't be achieved in isolation. It is a struggle that goes together with another model of society and of agriculture.

There are other aspects where we are also achieving this greater unity. I think that within La Vía Campesina there were also important steps in the organization and in the processes of articulation at the level of countries as well as at the level of large regions. We, as members, have to assume these principles and values that are defended in La Vía Campesina and apply them inside of our organizations as permanent work of advancing the organization and consciousness, so that our communities also are applying the concepts of food sovereignty.

I think that we have to continue constructing from our practice, respecting the differences that exist, but the principles should have merit for all of us wherever we are. It is a constant struggle as far as social organization, of building and of having political coherence. We're speaking of a process of organization, of raising awareness, which is a long process.

Also, if we have cultural differences, differences in practices, ultimately what are the principles that unify them? We have to permanently work so that they are linked in the meaning of a sustainable agriculture. What does it mean for us to develop agroecology in practice as an agriculture project that confronts the capitalist model of monocultivation?

It takes the participation of men and women to both establish the principles and to do the political training. These principles have to be clear so that they orient us and help us to do permanent work inside of our organizations, our countries, our regions. And so that, in fact, the cultural differences do not make us start to lose sight of our horizon of constructing another model of agriculture.

And most of all it demands this belief in the human being, this deep belief that another world is possible. But to make it be possible it is necessary that human beings, as social and political subjects, be the principal actors of this transformation. And it is necessary to make a huge investment in human beings themselves, in the interior world, the mindset, of the men and women who cultivate the countryside.

Note

1.   The Emmaus homeless non-profit organization was started in France in the 1940s and today operates in thirty-six countries. See <http://emmaus-international.org/>.

4

"Drawing Forth the Force that Slumbered in Peasants' Arms"
The *Economist*, High Agriculture & Selling Capitalism

Jim Handy & Carla Fehr

The idea of food sovereignty proposes a dramatic reorganization of the social relations of production within agriculture. It stems from several suppositions about the effect of current and past agriculture policies: by driving peasants and small farmers from the land, agricultural production is left to large, highly capital-ized farms, and agricultural markets function to benefit large-scale, monopolistic and long-distance trade. The result, food sovereignty proponents suggest, has not only been social and economic dislocation but increased fragility of the food pro-duction system and environmental degradation as agriculture comes to rely ever more heavily on chemical and petroleum inputs.

Those who defend current agricultural practice tend to disagree with these suppositions. They argue, instead, for further industrialization of agriculture, suggesting that the recent agricultural crisis necessitates increased scientific and capital investment in agriculture, more specialization, dramatically freer trade in agricultural goods and increased rural-urban migration as more peasants are pushed from the land (see Collier 2008; Trewavas 2008; Zoellick 2008; *Economist* 2008; World Bank 2008). To many sensible people, these latter propositions would seem nonsensical: one confronts a crisis in food production by driving more producers from the land, addresses the fragility of the current system of agricultural produc-tion by advocating more of the same and combats rural poverty by turning the rural poor into even poorer urban dwellers!

How are we to explain the continued faith in industrial agriculture? One of the reasons for its persistence is that for over two and a half centuries, since the middle of the eighteenth century, capitalists, politicians and classical liberal economists have determinedly proselytized for it, selling their faith in industrial agriculture to a sometimes skeptical public. They have been aided in this pursuit by academics and others who have sought to explain the emergence of Britain as an industrial power — and the wealth of its citizens — as a result of the changes in British agriculture from the mid eighteenth to the mid nineteenth centuries. The British experience has developed mythic qualities; elements of those agricultural changes have been fetishized and exported around the world. Perhaps the most fetishitic aspect of this extrapolation from the British experience has been the perceived need to push commoners, peasants and small farmers from the land, turn them into wage labourers; reserve land for educated, informed and modern farmers, who, most importantly, could apply capital to the land; and break up the "hard clods of ignorance, prejudice, sloth and indifference" that prevailed in the countryside, according to the *Economist* in 1843 (*Economist* 1843a: 27; see also Handy 2009).

That this view of the benefits of the British experience in agriculture should become so all pervasive, despite the fact that it contributed to a dramatic increase in poverty and social dislocation, is a function of the success of its proselytizers, who helped pave the way for the increased inequality and destitution that was the most obvious result of the emergence of capitalism. Perhaps foremost among these was the *Economist* newspaper, which began publication in 1843. In this chapter, we examine the rhetoric that accompanied the change in British agriculture between the mid eighteenth and mid nineteenth centuries, outline and explain the success of that rhetoric and link it to enclosures inherent in current neoliberal food and agricultural policies.

A Land without Peasants

One of the most pressing questions of modern social inquiry has been: Why did Europe, and especially Britain, become wealthy? Economists, historians and others who tried to answer this question have suggested a myriad of often confusing and contradictory arguments (for examples, see Jones 1981; Landes 1998). Most often, however, these explanations focused on changes to agriculture in Europe, especially changes that occurred in Britain between the late sixteenth and the middle of the nineteenth centuries (Diamond 1998; Fagan 2000; Bernstein 2004).

Beginning in the sixteenth century, agricultural land in England began to be "enclosed" into private property. Much of this land had been common or "waste" land used by villagers for pasturage or cultivation. A significant percentage had been village land, which was allotted to individual families for cultivation but was controlled by the village and reallocated every year or so, or it was "owned" by landlords, but long tenure contracts and custom ensured that tenants had relatively guaranteed access to land. Most of the land enclosed in the sixteenth and seventeenth centuries was converted to sheep grazing as wool became England's biggest export and as the English woolen textile industry developed.

After 1750, a new round of enclosures, focused on "improved cultivation," occurred. As enclosure continued, many landlords sought either to improve their own land through capital investment or to lease the land to new tenants who would pursue capitalist agricultural practices: invest capital in the land, seek the highest returns for their improvements and market their products. In the process, more commoners were forced from the land, and their "open" — so-called because they were often unfenced — fields were replaced by capitalist farms, to the extent that by the middle of nineteenth century, as Eric Hobsbawm and George Rude suggest, England "presented a unique and amazing spectacle to an enquiring foreigner; it had no peasants" (1968: 3).

Many analysts suggest that this process was an essential, perhaps *the* essential, element in fostering British prosperity and was central to the development of capitalism. The changes to British agriculture, they argue, led to dramatic increases in agricultural productivity, inspiring a host of other beneficial economic changes. Britain experienced marked improvements in food security even as a smaller percentage of the population was engaged in agriculture — in 1700, 82 percent of the British population lived in rural areas, in 1800 that percentage was 66, and by 1901 it had fallen to 24 percent (Woods 2000: 362). Agricultural production became more efficient and more scientific, ready to experiment with new techniques and crops and more geared to the market. Most essentially, this literature argues, changes to agriculture allowed a now surplus rural population to improve their standard of living by migrating to the cities to take up newly created industrial jobs in Britain's "high wage" economy (Fagan 2000; Bernstein 2004). Some even argue that the new agricultural practices — often called "high" agriculture — were especially environmentally sustainable, integrating a closed system of moderate pasturage, animal manure and environmental stewardship to agricultural practice (Duncan 1996: 54–55, 64–68, 95–97).

The Mythical Benefits of Enclosure

But how beneficial were the changes to British agriculture during this period? There is no doubt that some of the newly capitalized farms in England in the eighteenth and nineteenth centuries engaged in extensive works to try to improve land and agriculture production, from draining heavy land, planting new types of pasturage and growing new crops for both foodstuffs and animal feed. There also appears to be little doubt that England enjoyed almost unprecedented food security through much of this period; England suffered from no major famine after the middle of the seventeenth century. But, are these processes related?

Recent research suggests that enclosure did not lead to dramatic increases in agricultural productivity. In fact, agricultural productivity growth actually slowed down dramatically during the height of parliamentary enclosures after 1750 and didn't increase significantly until after 1800. Agricultural productivity increased significantly again for some decades following 1800, but this fairly quickly tapered off as the availability of fertilizer declined in the middle decades of the nineteenth century (Allen 2009: 59).

How can this be the case if this is also the period of the most dramatic growth in enclosed land and capitalist agriculture? The answer to that question seems to be that large, highly capitalized, enclosed estates were not significantly more productive than open fields of common land in England or than small peasant agriculture elsewhere. Robert Allen's careful assessment of increases in agricultural productivity in England indicates that open field commons were only significantly less productive in relation to labour or land in areas requiring the most investment in drainage, and nowhere else. Indeed, if total factor productivity is considered, that is, if returns to capital expenditures are taken into account, there was almost certainly little advantage. In addition, open field commoners were often as ready to experiment with new crops and new techniques as larger farmers. Indeed, Allen suggests that open fields might have been the ideal locale for such experimentation as the small size of each person's rows allowed for small changes in agricultural techniques or new crops while others in the region watched to see how the experiment turned out (Allen 2009: 60–74).

This fits well with what we know about peasant agriculture elsewhere. Peasant agriculturalists everywhere in the world have proven to be quick to adopt new plants and new techniques to improve agricultural production. It was not, primarily, large farmers who fostered the spread of New World agricultural products around the world, from Ireland to China. Rather, a host of new cultigens emerged from the Americas and remarkably quickly became peasant crops in locales all around the world. Jan Douwe van der Ploeg argues that Dutch agriculture was not particularly effective or efficient until after a period of "repeasantization" in the latter half of the nineteenth century, which allowed producers to develop a particular mix of labour intensive crops on small farms (2008: 47). Much the same is true for small-scale Flemish agricultural producers, who were derided by supporters of large-scale, capital-intensive agriculture in Britain (*Economist* 1851a).

The often-cited improvements in British food security and nutrition levels are also not so impressive when examined more closely. Through the eighteenth and nineteenth centuries, England continued to import a substantial portion of its food, in the form of wheat and meat, from the Baltic countries. During the latter part of the eighteenth century and into the nineteenth, as the Corn Laws restricted imports from elsewhere, Ireland became a major source of grain for England, providing hundreds of thousands of tons of wheat each year, including 300,000 tons in both 1846 and 1848, and only slightly less in 1847, in the midst of the Irish potato famine (Edwards 1973: 218; O'Grada 1994: 120; Ross 1998: 43–47).

Significantly, England imported this food even after it was able to free up a large portion of its land for the production of agricultural goods as it retooled the textile industry away from wool production. Increasingly, English textiles turned from wool that was produced domestically to linen (from flax, a significant proportion of which was grown in Ireland) and cotton (imported from the Americas or from colonies like India). Kenneth Pomeranz estimated that to replace the cotton imported to feed England's textile mills in the 1830s with wool produced domestically would have required 23,000,000 acres, significantly more than all the arable land available in England (2000: 274–76; also Goody 1996: 127). Even though England imported the raw material for textile production, saving hundreds of thousands of acres that could then be used for food production, an increasing percentage of the caloric intake of English workers was met through the consumption of sugar. Sugar, like cotton, was produced primarily from slave or indentured servant labour in colonies. By the end nineteenth century, sugar made up close to 20 percent of English calories, marking an important shift to the increased consumption of appetite appeasing stimulants (coffee, tea, tobacco, sugar) as the British diet worsened (Mintz 1986: 67). Any assessment of food security that also takes into consideration access to nutrition in those areas that "relieved" England might come to the conclusion that English food security in the eighteenth and nineteenth centuries grew not primarily through increased agricultural productivity in England but through dispossession elsewhere and the process of exporting famine (Davis 2001).

Were those forced from the countryside to cities to participate in England's industrial production better off, as has been argued? There is significant debate about changes in real wages in England through the eighteenth and nineteenth centuries. It seems that real wages in the cities and real incomes in the countryside were probably stagnant or declining until at least 1800, possibly 1820, and rose slowly after that (Marx 1967: 73–75; Pomeranz 2000; Allen 2001). After the first few decades of the nineteenth century, there was indeed a significant premium in urban wages, especially in the largest centres, over rural wages, supporting the argument that those who migrated to the cities were better off. But these numbers need to be treated with some suspicion. Income levels cannot be taken to indicate standard of living. There can be little doubt that urban workers were more poorly nourished than rural dwellers throughout this period. They might well have been more poorly nourished in the middle of the eighteenth century than they had been in the seventeenth century as well. There is some evidence that there was less diversity in foodstuffs and a reduced availability of dairy products and vegetables during the eighteenth century (Perelman 2000: 297; Neeson 1993: 23). In a situation in which slight increases in wages did not make up for reduced access to food, increased incomes — even if they did occur — could be said to have little beneficial impact on the standard of living.

Certainly, examinations of height and mortality rates during this period suggest that rural dwellers were significantly healthier than urban workers. Studies of the relative height of prisoners and conscripts to the British Army and the East India Company Army through the eighteenth and the nineteenth centuries show that English urban dwellers were the shortest group on average; shorter than rural conscripts and shorter even than Irish and Scottish urban conscripts despite the poverty apparent in Ireland. Moreover, the average height of recruits declined, both rural and urban but much more markedly among urban dwellers, inexorably if unevenly from 1740 for the next 120 years — that is, over the very years that changes in British agriculture were meant to have dramatically improved agricultural productivity and food security (Komlos 1993; Nicolas and Stekel 1991).

Mortality rates indicate similar and even more striking discrepancies, although nutrition would be only one element in determining differential mortality rates among urban and rural dwellers. In 1701, the average life expectancy in London was eighteen and a half years; in rural areas of England it was almost two and a half times that, over forty-one years. While over the next century and a half, life expectancy in London improved, it was by 1850 still almost ten years less than that in rural areas. Nor was London the worst case; the major industrial cities in the North of England — Liverpool, Manchester and Birmingham — lagged significantly behind. Average life expectancy in Liverpool in the 1850s was twenty-five years, just slightly more than half that of a rural dweller in England 150 years earlier (Woods 2000: 364–71).

While mortality rates, and even height, might have been affected as well by a deteriorating environment in urban areas as coal-fired industrialization took its toll, did high agriculture at least prompt better environmental stewardship in rural areas of England? The evidence suggests otherwise. If high agriculture was originally perceived to be a relatively benign, closed agricultural system utilizing livestock as mobile fertilizer, that vision was quickly abandoned. Instead, by the second quarter of the nineteenth century, the focus was on the use of imported nitrates, most prominently guano from the coast of Peru; by 1850, England was importing 116,295 tons a year. The application of guano, according to the *Economist*, one of the most ardent supporters of its use, helped distinguish high farming from farming that took "only the natural produce of the soil" and meant that "the growth of corn may be enlarged to an extent that for all practical purposes we may call unlimited" (*Economist* 1851b, 1847a, 1847b, 1848a, 1857). But guano, accumulated over thousands of years on the Peruvian coast, at this level of exploitation lasted only a few decades, and the brief rise in agricultural productivity it inspired in the second quarter of the nineteenth century soon ended (Pomeranz and Topik 1999: 126–29). Thus, rather than being environmentally benign, high farming actually initiated a practice of mining the soil and looking frantically elsewhere for similar artificial nitrate sources.

On virtually all levels, then, even a relatively brief examination of the evidence suggests that most of the arguments put forward about the beneficial impacts of the changes in British agriculture during this period are illusory. Why then should they have prompted such staunch defenders? To understand the enduring support for industrial agriculture we need to examine the determined campaigns launched by its proponents in eighteenth- and nineteenth-century Britain. Their work inspired a series of tropes about the benefits of capitalist relations of production in the countryside and of industrial, scientific or "high" farming, which was essential to the capitalist process; and corresponding to this a series of negative myths associated with its opposite: peasants and small-scale agricultural production.

"The Superstitious Worship of S's Name"

Those promoting enclosure and high agriculture in England in the late eighteenth and nineteenth centuries fostered two necessary myths: enclosure was needed for increased agricultural production and would benefit society in general, and the shift to a market society and capitalist relations of production happened through the natural consequences of the market, not through government intervention and oppression.

One of the prevailing concerns of many Enlightenment philosophers had been what would give order to society if distinctions based on birth and inheritance were eliminated. For a significant portion of them the answer lay in the almost magical workings of the market and the argument that the pursuit of individual self-interest in the context of a free market would lead to public wellbeing. Perhaps the first expression of this was in Bernard Mandeville's *Fable of the Bees, or Private Vices, Public Benefits,* written in 1714. But, the most celebrated exposition of this idea was, of course, Adam Smith's 1776 *An Inquiry into the Nature and Causes of the Wealth of Nations.* Smith argued that through the "invisible hand" provided by the market and "the private interests and passions of men [sic]," economic activity would be directed to pursuits "most agreeable to the interests of the whole society" (Smith 1809, vol. 2: 242, 495). This process would lead not only to unexpected social harmony but also to "that universal opulence which extends itself to the lowest ranks of the people" (Smith 1991: 10). Thus, the market was both natural and self-regulating, functioning "without any intervention of law" (Smith 1809, vol. 2: 495).

There were significant difficulties and contradictions in Smith's work. He provided no clear idea of how such a "natural" system selected some for menial labour, except some vague ideas about natural abilities. He also admitted that a division of labour such as he described would result in workers becoming "as stupid and ignorant as it is possible for a human creature to become" (Smith 1809, vol. 3: 194). While observing that employers constantly sought to keep wages low, he never successfully explained how increased population — or an increase in those seeking wages because they were thrown off the land — would not lead to falling wages rather than "universal opulence" (Smith 1991: 59). Partly because of these contradictions and a certain naivety inherent in the argument, the book was not, initially, highly regarded even by other writers important in fostering new ideas about economics.

By the first couple of decades of the nineteenth century, the faults of the book, given the character of the time, began to be seen as benefits. Principal among these was Smith's insistence that the functioning of the market was natural whenever people didn't pass laws that made "that a crime which nature never meant to be so" and that this process would benefit everyone — that free markets and factory labour would lead to "that universal opulence which extends itself to the lowest ranks of the people." It was in these two areas that Smith differed fundamentally from most other observers.

Since at least the early seventeenth century, one of the principal preoccupations of landlords and capitalists had been how to force people to work for them. They had almost universally agreed that it was only the gravest poverty that would compel people to work for others and that whenever they could relieve that poverty through working on the land for themselves, they would do so. It was here that the classical economists and those arguing for the enclosure of the commons were most perfectly in consort. Bernard Mandeville in the early 1700s, while on the one hand arguing for the "social harmony" that results from the pursuit of private vices, made it clear that those private vices depended on people driven to labour by poverty:

> It would be easier, where property is well secured, to live without money than without poor; for who would do the work…. Those that get their living by their daily labour… have nothing to stir them up but their wants which it is prudence to relieve, but folly to cure…. it is manifest, that in a free nation, where slaves are not allowed of, the surest wealth consists in a multitude of labouring poor… without them there could be no enjoyment and no product of any country could be valuable. (cited in Marx 1967: 614–15)

For the next 150 years, at least, most careful observers understood the need for real or imagined poverty in creating labourers. James Steuart, writing in 1767, observed that under slavery, "men were… forced to labour because they were slaves to others, men are now forced to labour because they are slaves to their own wants" (cited in Perelman 2000: 151). Or, as Arthur Young succinctly expressed it, "Everyone but an idiot knows that the lower classes must be kept poor or they will never be industrious" (Porter 2000: 377). The chief impulse to labour, they understood, was hunger. As the Reverend Joseph Townsend, writing near the end of the eighteenth century, expressed it, "Legal constraint is attended with too much trouble, violence and noise… whereas hunger is not only a peaceable, silent, unremitted pressure, but as the most natural motive to industry and labour, it calls forth the most powerful exertions" (cited in Marx 1967: 646–47).

These ideas were often used to attack the poor laws, which, it was felt, allowed people to shirk labour because they did not feel the full brunt of hunger. But even more pernicious than the poor laws in preventing hunger from working its magic was the access the rural poor had to land in the shape of commons and other customary tenure. The unwillingness of the English poor to work when they had access to land was noted as early as the seventeenth century. As one of Cromwell's advisors expressed it, enclosure "will give the poor an interest in toiling, whom terror never yet could enure to travail" (cited in Hill 1972: 52). A century and a half later, the Reverend John Howlett, a friend of Adam Smith and a chief justice, argued that enclosures were beneficial precisely because this would turn poor farmers into even poorer labourers and encourage population growth. "In particular," he said, "enclosure would provoke a rapid and general increase of labouring and then of indigent poor" (cited in Neeson 1993: 27). This would lower wages and benefit industry. Into the nineteenth century, as enclosures continued, commentators also continued to argue that the independence of the rural population needed to be reduced in order to force them to labour. As J.M. Neeson points out, complaints about the lack of labour were common: Charles Vancouver, writing in 1813, argued that rural wages were too high and that enclosure was needed to reduce the independence of the poor. Even when enclosure was complete, the country needed to be "vigilant" to prevent any reassertion of independence. Even the hedgerows needed to be chosen with care to insure they could not provide sustenance to the poor, who should be forced to live in cottages provided by the landlords to avoid the "corrupt solidarity of the village" (cited in Neeson 1993: 28–30).

It was this process of enclosure and the attendant changes to the poor laws, designed to create poor labourers from English commoners, that Karl Marx most closely detailed. As Marx made clear, this was neither a natural process nor one not accompanied by conflict:

> The process, therefore, that clears the way for the capitalist system, can be none other than the process which takes away from the labourer the possession of his means of production; a process that transforms on the one hand, the social means of subsistence and of production into capital, on the other, the immediate producers into wage labourers.... These new freedmen became sellers of themselves only after they had been robbed of all their own means of production.... And the history of this, their expropriation, is written in the annals of mankind in letters of blood and fire. (1967: 714–15)

As Michael Perelman points out, when Thomas Malthus signed out the copy of the 1784 edition of Smith's *Wealth* from his school library in 1789, he was only the third person to have done so. However, in the closing decade of the eighteenth and early decades of the nineteenth centuries, as concern about popular unrest in response to continued enclosure became heightened, what Adam Shapiro (1993: 53) calls the "depoliticization impetus of Smith's narrative" proved to be most useful. Smith's "comforting message" provided the basis for much of the argument for the reshaping of the British economy and English agriculture. As Perelman argues, "As a poet of economic harmony, Smith was second to none." He also points out that many of the classical economists understood both that Smith's vision was too rosy and that government intervention and repression were necessary for classical liberalism to work. He recounts how when Francis Horner, an editor of the *Edinburgh Review*, was asked to write notes for a new edition of *Wealth*, he refused, indicating, "I should be reluctant to expose S's errors before his work had operated its full effect. We owe much at present to the superstitious worship of S's name; and we must not impair that feeling, till the victory is complete" (Perelman 2000: 176).

Significant forces were brought to bear to ensure that victory. The powerful semi-official Board of Agriculture was at the forefront in attempts to portray enclosure as both beneficial to society as a whole and necessary for English prosperity. Arthur Young, the long-time secretary of the board, was perhaps the most important chronicler of the supposed benefits of enclosure. He engaged in a series of tours through Britain and Ireland in the second half of the eighteenth century and often portrayed enclosed fields as "civilization" and the commons as barbarism. In one district he reported that "the Goths and Vandals of open fields" threatened the "civilization of the enclosures" and commented on how when talking with commoners, "I seemed to have lost a century of time" (cited in Porter 2000: 309). John Sinclair, the president of the board, writing in 1803, was even more explicit. After news about further British victories abroad, he remarked:

> Why should we not attempt a campaign also against our great domestic foe. I mean the hitherto unconquered sterility of so large a proportion of the surface of the kingdom? … let us not be satisfied with the liberation of Egypt, or the subjugation of Malta, but let us subdue Finchley Common; let us conquer Hounslow Heath; let us compel Epping Forest to submit to the yoke of improvement. (cited in Turner 1984: 23)

Enclosers often portrayed the commons as of little use to the rural poor; in the words of one defender of the commons, the encloser must "bring himself to believe an absurdity, before he can induce himself to do a cruelty" (Neeson 1993: 38–39). But believing that seemed to be increasingly difficult by the early nineteenth century. Some of the staunchest proponents of enclosure — John Howlett and Arthur Young most notably — began to question the effects of the practice, and the powerful allotment movement began in the first few decades of the century to provide rural workers with plots of land to grow their own food. It was in this context that the *Economist* newspaper began its work in shoring up support for capitalist relations of production in the countryside and high agriculture.

"A Business to Be Undertaken by Capitalists"

The *Economist* was started in 1843 by James Wilson, a hatmaker who gained his reputation as a vocal opponent of the Corn Laws, which restricted trade in grains. He remained editor of the paper and wrote much of it for more than a decade before turning control over to his son-in-law, Walter Bagehot. By the mid 1860s, it was a respected and influential voice, shaping opinion among powerful sectors of society and the upper middle class on both sides of the Atlantic. Born in the struggle for free trade in agriculture, the *Economist* was staunchly, even vehemently "Smithsonian" in its adherence to classical liberal arguments, with its continued reiteration that such practice would benefit all — that is, all but the most undeserving.

The newspaper continually expressed its opposition to anything that might stand in the way of the "natural operations" of the marketplace. As the editor argued, "The more we give or allow scope to the free exercise of self-love, the more complete will be social order" (*Economist* 1848b). The most essential component in that process, the paper suggested, was to prevent government regulation from interfering. In 1844, it warned, "The great practical lesson, which society has at present to learn, is that our greatest social inconveniences, though caused by laws, are to be cured only by an utter absence of legislation" (1844a, 1844b). The *Economist* made these arguments using rhetoric and a writing style that suggested that its proposed policy was the only logical one. It did this most successfully in its careful use of statistics. The science of statistics, the paper said, would do more "to wipe away those intricate masses of cobwebs which narrowed prejudices, unworthy jealousies, unchristian animosities and inflamed passion had contrived" (1843b).

Like Smith himself, though, the paper was full of contradictions and, despite its appeal to elevate public discourse, often extremely mean spirited. It definitely defended government regulations that favoured capital; it was an ardent supporter of enclosures, which, after all, were accomplished through acts of Parliament, and it was decidedly in favour of government interference when it was needed to curtail worker and peasant organization and opposition. Like so many liberal economists before and after this period, the *Economist* was only opposed to government interventions it considered troublesome. The most troublesome was anything that prevented a particular type of change in the countryside.

Unlike Adam Smith, who, while counselling its dissolution, believed that rural society produced the most intelligent and nimble inhabitants, the *Economist* argued that the countryside was rife with ignorance, prejudice, nonsensical ideas and lazy people. Farmers were content to let society progress around them, while those who were still employed in the cultivation of the soil were "avowedly the most wretched and ignorant portion of [the] population" (1844c, 1844d). This indolence, it argued, was the result of depending on the benevolence of the land. Thus, the most fertile and abundant natural environments created the most dissolute people. For the *Economist*, therefore, it was in cities where the great advances in society were occurring, in cities where capital and science and progress were wedded. But, rural areas could progress to the extent that they abandoned old habits and adopted new models (1853a).

The *Economist*'s new model for English agriculture was of a particular type; it was now "a business to be undertaken by capitalists" (1848c). The first requirement was the dominance of capital. It was capital that would transform English agriculture, capital that would cultivate the soil and capital that would ensure English agriculture was progressive. The *Economist* assured readers that the "great secret of farming we think is this — the judicious application of a sufficient amount of capital to the soil..." (1849). Indeed, the paper argued, if there were not men of capital to invest in the land it should not be cultivated. In this regard, the *Economist* was not unique but rather reflected a generalized fascination with the purported magic of capital. In 1821, an Englishman, writing under the pseudonym of Piercy Ravenstone, described the temper of the age in this way:

> Where reason fails, where argument is insufficient, it operates like a talisman to silence all doubts…. It is… the great mother of all things, it is the cause of every event that happens in the world. Capital, according to them, is the parent of all industry, the forerunner of all improvements. It builds our towns, it cultivates our fields, it restrains the vagrant waters of our rivers, it covers our barren mountains with timber, it converts our deserts into gardens, it bids fertility arise where all before was desolation. It is the deity of their idolatry which they have set up to worship in the high places of the Lord; and were its powers what they imagine, it would not be unworthy of their adoration. (cited in Pasquino 1991: 105–106)

For the *Economist,* the triumph of capital required that land be treated like any other commodity, freely disposed of on the market without the "costly and obsolete rubbish by which… its free disposal is cruelly hampered" (1853b). For this to occur, the primary requisite was the complete victory of enclosure. All common rights originated in uncivilized times, the *Economist* argued, when the population was thin and when land was not appropriated, or when no one troubled himself that others used his property — that property being worthless, until such use had been converted into a legal right (1845, 1847c). Enclosure was in the process of freeing agriculture from these bounds. Indeed, the beneficial results were so obvious, the paper remarked, "that it is surprising to find intelligent men disputing the advantages of enclosing commons" (1845).

But not all kinds of private property in land were equal, nor, according to the *Economist,* would they all contribute to progress. The paper consistently argued that land needed to be held in large estates run according to scientific principles. It was never able to provide a clear explanation, despite trying on a number of occasions, why such large estates were necessary. For example, it argued that farmers with only limited amounts of capital should have the size of their holdings reduced so as to maximize the capital available to each quantum of land. Carried to its logical extension, this argument would suggest that all large estates would do better if some land were taken from them. Mostly, the *Economist* argued for size by denouncing small, suggesting that the idea that small farms, even when cultivated with "great industry," could be as productive as large capitalist estates was a "fallacious notion … everywhere contradicted by facts and experience. Usually petit-farming is a miserable affair" (1851a).

It was in the treatment of rural labourers and commoners made superfluous by the transformation of farming that the *Economist* was most draconian and most contradictory. In response to concerns expressed by some about reduced rural employment, the paper argued that the employment of capital meant more labour was needed on farms. In the face of clear evidence to the contrary, the paper clarified its stance: the transition to capitalist agriculture increased the employ of qualified and responsible, permanent workers. Agricultural change had "drawn forth the force which slumbered in the peasants' arm; the result has been that, though the labourers are fewer, they have done more work than heretofore" (1861a). In other words, those set adrift by enclosure and the consolidation of land were commonly the feeblest and least effective, men of "dissolute and unsteady habits... very commonly consum[ing] with utter improvidence the large wages they earn[ed] during the summer months, and [gone] into the union workhouse during the winter." They were greedy and drank too much but, the paper maintained, with mowing and reaping machines, horse rakes, hay-making machines and the like, farmers could discard the services of Irish reapers, militiamen and other half-labourer, half-vagabond assistants on whose aid they had been too dependent (1851c).

When other powerful sectors of society began to express concerns about the nature of the changes in rural England and to search for ways to ameliorate the suffering caused there, the *Economist* was steadfast and sought to marshall spirits for the battle ahead. When faced with a constant flow of impoverished migrants to the cities, the *Economist* was undeterred and blamed "the deluge of pauperism which is now flowing in upon the towns from the country" on "the feudal distribution of the land" rather than the expulsion of the rural population through enclosure (1844e). Indeed, it embraced that transformation, arguing that the decreased proportion of the people devoted to agricultural pursuits had already "broken down the parochial and patriarchal barriers which made each spot of the land a gaol, though a home, for a particular portion of the community, and the same progress will cause them to be entirely removed" (1855).

In the 1840s, a movement began to contemplate the provision of small rural allotments to those thrown off the land. The Parliamentary Select Committee on Allotments argued: "The allotment system... appears to be the natural remedy for one of the detrimental changes in the condition of the labouring classes of this country... throwing them for subsistence wholly and exclusively upon wages" (Moselle 1995: 482–83). In the face of such backsliding, the *Economist* went on the attack. Agreeing with such notable defenders of the poor as Thomas Malthus — who from his perch as professor of political economy at the East India Company University, argued that it would lead to a catastrophic increase in population (Burchardt 1997: 166; Ross 1998: 1) — the paper argued that dividing the land and establishing people, without capital, on small farms was an "irrational custom" founded upon "Irish ideas" (1848d, 1848e, 1848f). Allotments, with their "character of stern utility, of substitute for wages, of last source of subsistence," were evidence of "the deepest poverty and the last stage of dependence." The real sin, however, was that the labourer would reserve some of his strength for the cultivation of his allotment, and thus his employer would not "have all the sweat of his brow" (1844f, 1844g, 1844h).

It was not, the paper often assured people, that it had no sympathy for the rural poor forced out of the countryside. Rather, it was convinced that "the more you help people, the poorer and more dependent they grow" (1859a, 1859b). They needed to get to the city, find a job and prosper. These were the kinds of "most elementary and universally acknowledged truths" of which it was sometimes necessary to remind the public (1850).

From Liberalism to Neoliberalism

Pierre Bourdieu describes neoliberalism as "a program of the methodical destruction of collectives" (cited in Davis and Monk 2007: x). This aptly describes eighteenth- and nineteenth-century liberalism as well, especially in the extension of agrarian capitalism. In 1861, when William Trotter, a member of a local farmers' club, expressed reservations to the *Economist* about the tendency, which seemed to typify the age, "of collecting wealth into heaps, and population into dense masses," the paper assured him that this was just the "natural tendency" of increased wealth and prosperity and a "sign of general progress" (1861b). There is no indication whether Trotter was reassured by the explanation. It is clear, however, that many people of power and influence were. Despite the negligible or questionable benefits in increased productivity of land and labour such destruction entailed, the basic precepts of high agriculture continued to be extolled and exported for the next century and a half.

Enclosure continues in our time, its benefits extolled by the same actors, including the *Economist*. When the World Bank argued in its 2008 *World Development Report* that continued rural-to-urban migration was a good thing and should be embraced, the *Economist* enthusiastically agreed. In 2009, in the midst of mounting opposition to huge purchases of land in "poor" countries by wealthier nations and corporations in those nations, the *Economist* raised concerns. But, its editors were not disturbed by the concept itself, which they endorsed. Rather, they took issue with the fact that these purchases had not occurred transparently on the open market and that those purchasing the land were not sufficiently committed to "reorganizing the way people work" (2008, 2009).

Modern agriculture was based on a set of exclusions and enclosures that were fundamental to the emergence and strengthening of capitalism. Through the eighteenth and nineteenth centuries, a set of myths about the supposed benefits of capitalist agriculture were constructed and continually reinforced to help make these exclusions more palatable. Food sovereignty challenges these myths in important ways, because it demands that we rethink what was at the very centre of that transition; it demands that we treat food not simply as a good, access to which and the production of which is determined by a mythically natural and fetishized "market." Food sovereignty demands that we recognize the social connections inherent in producing food, consuming food and sharing food. In the process it will change everything.

References

Allen, R. 2009. *The British Industrial Revolution in Global Perspective*. Cambridge, UK: Cambridge University Press.

_____. 2001. "The Great Divergence in European Wages and Prices from the Middle Ages to the First World War." *Explorations in Economic History* 38.

Bernstein, W. 2004. *The Birth of Plenty: How the Prosperity of the Modern World Was Created*. New York: McGraw-Hill.

Burchardt, J. 1997. "Rural Social Relations, 1830–50: Opposition to Allotments for Labourers." *The Agricultural History Review* 45, 2.

Collier, P. 2008. "The Politics of Hunger." *Foreign Affairs* 87, 6 (December).

Davis, M. 2001. *Late Victorian Holocausts*. London: Verso.

Davis, M., and D.B. Monk (eds.). 2007. "Introduction." In *Evil Paradises: DreamWorlds of Neo-Liberalism*. New York: New Press.

Diamond, J. 1998. *Guns, Germs, and Steel: The Fate of Human Societies*. New York: W.W. Norton and Company.

Duncan, C. 1996. *The Centrality of Agriculture: Between Humankind and the Rest of Nature*. Montreal: McGill-Queen's Press.

Economist. 1843a. "Scientific Agriculture for Farmers." September 3.

_____. 1843b. "Introductory Remarks, Explanatory of the Following Tables." Supplement, November 4.

_____. 1844a. "Colonel Torrens on a Ten-Hours Bill." April 20.

_____. 1844b. "Intolerance of Free Traders." December 28.

_____. 1844c. "Proposed College of Chemistry." August 3.

_____. 1844d. "The Cultivation of Waste and Other Lands. Native Industry." December 28.

_____. 1844e. "Incendiary Fires in the Country." January 13.

_____. 1844f. "Division of Labour: The Allotment System." November 23.

_____. 1844g. Letter to the Editor: "Mr Greg on Allotments." November 23.

_____. 1844h. "The Agricultural Labourer and the Regulations Imposed in the Allotment System." December 21.

_____. 1845. "Enclosure of Waste Lands: Parliamentary Evidence on the Subject." April 26.

_____. 1847a. "Obstacles to High Farming." February 6.

_____. 1847b. "The Progress of Husbandry in England." July 31.

_____. 1847c. "Property in Land-Tenant-Right." November 27.

_____. 1848a. "High Farming Profitable." November 11.

_____. 1848b. "New Means and New Maxims." April 22.

_____. 1848c. "Agricultural Customs Committee." September 16.

_____. 1848d. "Laws of Succession to Property." January 22.

_____. 1848e. "Large or Small Farms." February 12.

_____. 1848f. "Allotment Farming in Hertfordshire." June 24.

_____. 1849. Citing a story from the *Mark Lane Express*/ December 15.

_____. 1850. "Removal of Protection Not the Cause of Agricultural Distress." August 24.

_____. 1851a. "Flemish Farming." February 1.

_____. 1851b. "Guano — Adulteration." September 6.

_____. 1851c. "Cheap Food and Rural Wages." September 6.

_____. 1853a. "A Peculiarity of Agriculturalists." September 3.

_____. 1853b. "Real Property Law Amendment." February 5.

_____. 1855. "Scarcity of Labour." September 8.

_____. 1857. "Meat Supplies." July 18.

_____. 1859a. "The Condition of Rural Labourers." June 11.

_____. 1859b. "Dwellings of the Rural Labourers." September 3.

_____. 1861a. "The Agricultural Population." June 22.

_____. 1861b. "Good Farming: The Helps and Hindrances." May 4.

_____. 2008. "Lump Together and Like It." November 8.

_____. 2009. "Cornering foreign fields." May 21.

Edwards, R.D. 1973. *An Atlas of Irish History*. London: Methuen.

Fagan, B. 2000. *The Little Ice Age*. New York: Basic Books.

Goody, J. 1996. *The East in the West*. Cambridge, UK: Cambridge University Press.

Handy, J. 2009. "'Almost Idiotic Wretchedness': A Long History of Blaming Peasants." *Journal of Peasant Studies* 36, 2 (April).

Hill, C. 1972. *The World Turned Upside Down: Radical Ideas during the English Revolution*. London: Temple Smith.

Hobsbawm, E., and G. Rude. 1968. *Captain Swing*. New York: Pantheon Books.

Jones, E. 1981. *The European Miracle*. Cambridge, UK: Cambridge University Press.

Komlos, J. 1993. "The Secular Trend in the Biological Standard of Living in the United Kingdom, 1730–1860." *Economic History Review* (New Series) 46, 1 (February).

Landes, D. 1998. *The Wealth and Poverty of Nations*. New York: W.W. Norton.

Marx, K. 1967. *Capital, Volume 1,* translated from the 3rd German Edition, by Samuel Moore and Edward Aveling. New York: International Publishers.

Mintz, S. 1986. *Sweetness and Power: The Place of Sugar in Modern History*. London: Penguin.

Moselle, B. 1995. "Allotments, Enclosure, and Proletarianization in Early Nineteenth-Century Southern England." *Economic History Review* XLVIII, 3.

Neeson, J.M. 1993. *Commoners: Common Rights, Enclosures and Social Change in England, 1700–1820*. New York: Cambridge University Press.

Nicholas, S., and R.H. Steckel. 1991. "Heights and Living Standards of English Workers During the Early Years of Industrialization, 1770–1815." *The Journal of Economic History* 5 (December).

O'Grada, C. 1994. *Ireland: A New Economic History, 1780–1939*. Oxford: Clarendon Press.

Pasquino, P. 1991. "Theatrum Politicum: The Genealogy of Capital — Police and the State of Property." In G. Burchell, C. Gordon and P. Miller (eds.), *The Foucault Effect: Studies in Governmentality*. Chicago: University of Chicago Press.

Perelman, M. 2000. *The Invention of Capitalism*. Durham: Duke University Press.

Pomeranz, K. 2000. *The Great Divergence: China, Europe, and the Making of the Modern World Economy*. Princeton: Princeton University Press.

Pomeranz, K., and S. Topik. 1999. *The World that Trade Created*. London: M.E. Sharpe.

Porter, R. 2000. *The Creation of the Modern World*. London: W.W. Norton.

Ross, E. 1998. *The Malthus Factor: Poverty, Politics and Population in Capitalist Development*. London: Zed Books.

Shapiro, M. 1993. *Reading Adam Smith*. London: Sage Publications.

Smith, A. 1809. *An Inquiry into the Nature of the Causes of the Wealth of Nations*. 3 vols. Edinburgh: Mandall, Doug and Stevenson.

Smith, A. 1991. *The Wealth of Nations* New York: Alfred A. Knopf.

Trewavas, A. 2008. "The Cult of the Amateur in Agriculture Threatens Food Security." *Trends in Biotechnology* 26, 9.

Turner, M. 1984. *Enclosures in Britain, 1750–1830.* London: Macmillan.

Woods, R. 2000. *The Demography of Victorian England and Wales.* Cambridge, UK: Cambridge University Press.

van der Ploeg, J.D. 2008. *The New Peasantries: Struggles for Autonomy and Sustainability in an Era of Empire and Globalization.* London: Earthscan.

World Bank. 2008. *World Bank Development Report 2008: Agriculture for Development.* Washington, DC.

Zoellick, R. 2008. "A Ten Point Plan for Tackling the Food Crisis." *Financial Times* May 29.

5

Capitalist Agriculture, the Food Price Crisis & Peasant Resistance

Walden Bello & Mara Baviera

In 2006–08, food shortages became a global reality, with commodity prices spiralling beyond the reach of vast numbers of people. International agencies were caught flat-footed, with the World Food Program warning that its rapidly diminishing food stocks might not be able to deal with the emergency. Owing to surging prices of rice, wheat and vegetable oils, the food import bills of the least developed countries (LDCs) climbed by 37 percent from 2007 to 2008, from $17.9 to $24.6 million, after having risen by 30 percent in 2006. By the end of 2008, the United Nations (2009: 8) reported: "The annual food import basket in LDCs cost more than three times that of 2000, not because of the increased volume of food imports but as the result of rising food prices." These developments added 75 million to the ranks of the hungry and drove an estimated 125 million people in developing countries into extreme poverty (FAO 2008). Alarmed by massive global demand, countries like China and Argentina resorted to imposing taxes or quotas on their rice and wheat exports to avert local shortages. Rice exports were simply banned in Cambodia, Egypt, India, Indonesia and Vietnam. South-South solidarity, fragile in the best of times, crumbled, becoming part of the collateral damage of the crisis.

For some countries, the food crisis was the proverbial straw that broke the camel's back. Some thirty countries experienced violent popular actions against rising prices in 2007 and 2008, among them Bangladesh, Burkina Faso, Cameroon, Cote'd'Ivoire, Egypt, Guinea, India, Indonesia, Mauretania, Mexico, Morocco, Mozambique, Senegal, Somalia, Uzbekistan and Yemen. Across the continents, people came out in the thousands against uncontrolled rises in the price of staple goods their countries had to import owing to insufficient production. Scores of people died in these demonstrations of popular anger.

The most dramatic developments transpired in Haiti. With 80 percent of the population subsisting on less than two dollars a day, the doubling of the price of rice in the first four months of 2008 led to "hunger so tortuous that it felt like [people's] stomachs were being eaten away by bleach or battery acid" (Lindsay 2008: 46). Widespread rioting broke out that only ended when the Senate fired the prime minister. In their intensity, the Haiti riots reminded observers of the anti-IMF riots in Venezuela — the so-called Caracazo — almost two decades before, which reshaped the contours of that country's politics.

This chapter examines how neoliberal agricultural policies, particularly structural adjustment and trade liberalization, contributed to the global food crisis that became front-page news. We argue that the roots of the current food crisis lie in a fundamental transformation of agriculture promoted by corporate interests and international institutions. As corporate agriculture expands its reach and threatens peasant production, food wars and peasant protests are on the rise, struggles centred in repeasantization and food sovereignty.

A Perfect Storm?

The international press and some academics proclaimed the end of the era of cheap food, and they traced the origin to a variety of causes: the failure of the poorer countries to develop their agricultural sectors, strains on the international food supply created by dietary changes in China and India's expanding middle classes, who were eating more meat, speculation in commodity futures, the conversion of farmland into urban real estate, climate change and the diversion of corn and sugarcane production from food to agrofuels.

The United Nations (2009) spoke about the crisis being the product of a "perfect storm," or an explosive conjunction of different developments. According to the United Nations, speculative movements that brought about the global financial crisis in the summer of 2007 were implicated in the food crisis, as speculation by financial investors in commodities and commodity futures markets had a considerable impact on food prices. It could be argued, said the report,

> that increased global liquidity and financial innovation has also led to increased speculation in commodity markets. Conversely, the financial crisis contributed to the slide in commodity prices from mid-2008 as financial investors withdrew from commodity markets and, in addition, the United States dollar appreciated as part of the process of the de-leveraging of financial institutions in the major economies. (2009: 46)

Others, like Peter Wahl of the German advocacy organization World Economy Ecology and Development (WEED), were more emphatic, claiming that speculation in agro-commodity futures was the key factor in the extraordinary rise in the prices of food commodities in 2007 and 2008. With the real estate bubble bursting in 2007 and trading in mortgage-based securities and other derivatives collapsing, hedge funds and other speculative agents, they asserted, moved into speculation in commodity futures, causing a sharp increase in trades and contracts accompanied by little or no increase in agricultural production. It was this move into commodity futures for quick profits, followed by a move out after the commodities bubble burst, that triggered the rise in the FAO food price index by 71 percent during only fifteen months, between the end of 2006 and March 2008, and its falling back after July 2008 to the level of 2006 (Wahl 2009).

The Agrofuel Factor

While speculation was certainly among the factors that created a "perfect storm" in 2006–08, an even more prominent explanation was the diverting of cereals, especially corn, from serving as food to being used as agrofuel. On July 3, 2008, the *Guardian* exposed a secret World Bank report that claimed that U.S. and E.U. agrofuels policies were responsible for three quarters of the 140 percent increase in food prices between 2002 and February 2008 (Chakrabortty 2008). This figure was significantly higher than the 3 percent previously reported by the U.S. Department of Agriculture (USDA), Oxfam's estimate of around 30 percent, the IMF figure of 20 to 30 percent and the Organization for Economic Cooperation and Development's (OECD) 60 percent. The report's conclusion was straightforward:

> The most important factor [in the food price increases] was the large increase in biofuels production in the U.S. and the E.U. Without these increases, global wheat and maize stocks would not have declined appreciably, oilseed prices would not have tripled, and price increases due to other factors, such as droughts, would have been more moderate. Recent export bans and specula-tive activities would probably not have occurred because they were largely responses to rising prices. (Mitchell 2008)

Completed as early as April 2008, the Mitchell report — after Donald Mitchell, lead economist of the World Bank research team — was allegedly suppressed by the World Bank out of fear of embarrassing former U.S. President George Bush and his aggressive agrofuels policy (Chakrabortty 2008).

The agrofuel factor affected mainly U.S. farming, where much corn production was shifted from food to agrofuel feedstock. This is hardly surprising since over its last few years the Bush administration's generous subsidies, made in the name of combating climate change, made conversion of corn into agrofuel feedstock instead of food very profitable. Pushed by a corporate alliance that included some of the biggest names in the energy and agrifood industries, such as Exxonmobil, Archer Daniels Midland and Cargill, Bush made agrofuel development one of the pillars of his administration's energy policy, with the announced goal that renew-able sources should comprise a minimum of 20 percent of the energy portfolio in the transport sector within ten years.

In 2007, with the administration's active lobbying, the U.S. Congress passed the *Energy Independence and Security Act*, which focused on promoting agrofuels and the automobile fuel industry. The Act targeted a more than eightfold increase in agrofuels production, from 4.7 billion gallons in 2007 to at least 36 billion gal-lons in 2022 — unusually high standards that would evoke significant changes in agricultural production. As of late 2007, there were 135 ethanol refineries in opera-tion and seventy-four more being built or expanded (APEC 2008). Midwestern America saw itself slowly being transformed into a giant agrofuels factory. In 2008, around 30 percent of corn was allocated for ethanol, with rapid increases occurring since 2006. Not surprisingly, the strong mandate and generous subsidies, as well as high tariffs against imported sugar-based Brazilian ethanol, ensured that by 2008, around 30 percent of U.S. corn was being allocated for agrofuel feedstock, with a not inconsiderable impact on grain prices.

While the actual impact of agrofuel production was bad enough, the future impact in developing countries was even more worrisome. Huge land lease deals are said to be taking place within land-rich countries like the Philippines, Cambodia and Madagascar (AFP 2009). There are widespread reports in the international media of private firms and governments from countries that lack arable land striking lease agreements. Some of these deals are for food production, others for agrofuels, but with land being commodified, what is produced on the leased lands will ultimately depend on what is most profitable to bring to the global market at a given time. The most notable of these deals is Korean firm Daewoo Logistics' plan to buy a ninety-nine-year lease on over a million hectares of land in Madagascar for agrofuel production. Maize and palm oil will be cultivated on almost half of the arable land in the country (Spencer 2008).

Similarly, Cambodia and the Philippines are negotiating "agricultural investment" projects. Kuwait is trading loans for Cambodian produce. The Philippines and Qatar are currently negotiating the lease of 100,000 hectares of land. In effect, the food and energy crises are causing countries to secure food supplies and agrofuel feedstock in unconventional ways. It is no longer sufficient to import grains. The land that produces that grain must be secured through contracts. Land is now the desired commodity, to the detriment of local populations, who depend on the land for their own food consumption. Political elites in land-rich countries appear to be all too happy to oblige at the expense of their own country's food security — not surprising, since multimillion-dollar leases, such as that being offered by the Chinese to the Philippine corporate groups, are a strong incentive.

The World Bank, often on the wrong side of things, is astute in the case of agrofuels. What it says is likely to happen in the Philippines may well apply to other countries:

> The expected increase in demand for sugar for ethanol production, abetted by the incentives provided by the Biofuel Act which, among others, mandates a minimum amount of bio-fuel use, can be expected to further pull sugar prices up, and consequently food prices. In addition, the bio-fuel policy will most likely put pressure on extending cultivation on marginal lands and converting forests to agricultural uses, thereby worsening the impact of agriculture on natural resources. Finally, it is also expected that this policy will increase both the value of sugarcane farms and the difficulty in completing the Comprehensive Agrarian Reform Program in these areas. (2008a: 78)

Structural Adjustment & Trade Liberalization

While speculation on commodity futures and the expansion of agrofuel production have been important factors contributing to the food price crisis, long-term
processes of a structural kind were perhaps even more central. These factors, in
the years leading up to the food price spike of 2006–08, resulted in demand for
basic grains — rice, wheat, barley, maize and soybeans — exceeding production,
with stocks falling to 40 percent of their 1998–99 levels and the stocks-to-use ratio
reaching record lows for total grains and multiyear lows for maize and vegetable oils
(United Nations 2009). A key reason that "production has fallen woefully short of
growth in food demand," asserts the United Nations, was the degradation of the
agricultural sectors of developing countries owing to the marked "weakening [of]
investment and agricultural support measures in developing countries, resulting
in a condition in which productivity growth for major food crops has stalled, and
there has been no significant increase in the use of cultivated land" (2009: 48). As
a result of supply constraints resulting from lack of investment, the FAO reported,
"even before the recent surge in food prices, worrisome long-term trends towards
increasing hunger were already apparent," with 848 million people suffering from
chronic hunger in 2003–2005, an increase of six million from the 1990–92 figure
of nearly 842 million (2008: 3).

In short, there were conjunctural, structural and policy ingredients in the
mix that led to the food price spike, and certainly a key element was the massive
economic reorientation known as "structural adjustment." The structural adjustment programs, which were imposed by the World Bank and IMF on over ninety
developing and transitional economies over a twenty-year period beginning in the
early 1980s, were most likely the condition *sine qua non* for the global food price
crisis. A brief look at structurally adjusted agriculture in Mexico, the Philippines
and Africa provides ample evidence of the devastating impact of these programs.

Eroding the Mexican Countryside

When tens of thousands of people staged demonstrations in Mexico early in 2007
to protest a sharp increase of over 60 percent in the price of *tortilla*, the flat unleav-
ened bread that is Mexico's staple, many analysts pointed to agrofuels as the culprit.
Mexico had become dependent on imports of corn from the United States, where
subsidies were skewing corn cultivation towards agrofuel production. However,
an intriguing question escaped many observers: How on earth did Mexicans,
who live in the land where corn was first domesticated, become "dependent" on
imports of U.S. corn?

The Mexican food crisis cannot be fully understood without taking into account the fact that in the years preceding the *tortilla* crisis, the homeland of corn
had been converted to a corn importing economy by free market policies promoted
by the IMF, the World Bank and Washington. The food price crisis in Mexico must

be seen as one element in the concatenation of crises that have rocked that country over the last three decades and brought it to the verge of being a "failed state." The key link between the food crisis, the drug wars and the massive migration to the North has been structural adjustment.

In the countryside, structural adjustment meant the gutting of the government programs and institutions that had been built from the 1940s to the 1970s to service the agrarian sector and contain the peasantry. The sharp reduction or elimination of such supports had a detrimental effect on agricultural production and productivity. The capacity of peasant agriculture was further eroded by the program of unilateral liberalization of agricultural trade in the 1980s and the North American Free Trade Agreement (NAFTA) in the mid 1990s, which converted the land that domesticated corn into an importer of the cereal and consolidated the country's status as a net food importer.

The negative effects of structural adjustment and NAFTA-imposed trade liberalization were compounded by the halting of the five-decade-long agrarian reform process as the neoliberals at the helm of the Mexican state sought to reprivatize land, hoping to increase agricultural efficiency by expelling what they felt was an excess agrarian population of fifteen million people (Bartra 2004). The early 1990s saw the amendment of Article 27 of the Mexican Constitution, which institutionalized sweeping agrarian reform through the formation of communally owned property called *ejidos,* effectively protecting these from market forces. This constitutional change was accompanied by a slew of laws aimed at promoting the privatization of communal property to encourage large-scale investment in agriculture, including foreign direct investment.

The combination of structural adjustment and trade liberalization has created a superfluous peasant population driven by increased mass poverty to the cities, where employment opportunities are scarce, or out of the country altogether, to the United States. Over twenty-five years after the beginning of structural adjustment in the early eighties, Mexico is in a state of acute food insecurity, permanent economic crisis, political instability and uncontrolled criminal activity.

Creating a Rice Crisis in the Philippines

Like Mexico in the case of corn, the Philippines hit the headlines early in 2008 for its massive deficit in rice. From being a net food exporter, the country had become a net food importer by the mid 1990s, and the essential reason was the same as in Mexico — that is, the subjugation of the country to a structural adjustment program that was one of the first in the developing world. The program involved a massive reduction of funding for rural programs ranging from credit support to subsidies for fertilizers and pesticides, which had been set up during the Marcos dictatorship in an effort to convert the peasantry into a pillar of the regime.

The deleterious effects of structural adjustment, which sought to channel the country's financial resources to the repayment of the foreign debt, were compounded by the entry of the country into the World Trade Organization in the mid 1990s. This required that it end the quotas on all agricultural imports, except for rice. In one commodity after another, Filipino producers were displaced by imports. Contributing to the decline of agricultural productivity was the grinding to a halt of the agrarian reform program geared towards land redistribution, which was not only successfully stymied by landlords but was not accompanied by an effective program of support services such as those that aided successful land reforms in Taiwan and Korea in the 1950s and 1960s.

Today, the status of the Philippines as a permanent importer of rice and a net food importer is implicitly accepted by a government that does not see agriculture playing a key role in the country's economic development, except perhaps to serve as a site for plantations rented out to foreign interests to produce agrofuels and food dedicated for export.

Destroying African Agriculture

As a continent that imports some 25 percent of the food it consumes, Africa has been at the centre of the international food price crisis. Although Africa's massive food deficit is commonly attributed to the fact that the continent has not undergone the Green Revolution, which Asia and Latin America experienced, other external pressures were the cause of Africa's predicament.

As in Mexico and the Philippines, structural adjustment, with its gutting of government budgets — especially its drastic reduction or elimination of fertilizer subsidies — was the key factor that turned relatively underpopulated Africa from a net food exporter in the 1960s to the chronic net food importer it is today. As in Mexico and the Philippines, the aim of adjustment in Africa was to make the continent's economies "more efficient" while at the same time pushing them to export-oriented agricultural production in order to acquire the foreign exchange necessary to service their burgeoning foreign debts.

This doctrinaire solution, micro-managed by the World Bank and the IMF, intensified poverty and inequality and led to significant erosion of the continent's agricultural and industrial productive capacity. In Malawi it led, earlier this decade, to famine, which was only banished when the country's government reinstituted fertilizer subsidies.

Again as in Mexico and the Philippines, the right hook of structural adjustment was followed by the left hook of trade liberalization in the context of unequal global trading rules. Cattle growers in Southern and West Africa were driven out of business by the dumping of subsidized beef from the E.U., while cotton growers in West Africa were displaced from world markets by highly subsidized U.S. cotton. The World Bank now admits that by pushing for the defunding of government programs, its policies helped erode the productive capacity of agriculture:

Structural adjustment in the 1980s dismantled the elaborate system of public agencies that provided farmers with access to land, credit, insurance inputs, and cooperative organization. The expectation was that removing the state would free the market for private actors to take over these functions — reducing their costs, improving their quality, and eliminating their regressive bias. Too often, that didn't happen. In some places, the state's withdrawal was tentative at best, limiting private entry. Elsewhere, the private sector emerged only slowly and partially—mainly serving commercial farmers but leaving smallholders exposed to extensive market failures, high transaction costs and risks, and service gaps. Incomplete markets and institutional gaps impose huge costs in forgone growth and welfare losses for smallholders, threatening their competitiveness and, in many cases, their survival. (2008b: 138)

Rather than allow Africans to come out with indigenous solutions to the continent's agrarian crisis, however, the World Bank is currently promoting a development strategy that relies on large-scale corporate agriculture along with "protected" reserves where marginalized populations would eke out an existence based on smallholder and communal agriculture, for which the Bank does not see much of a future (Havnevik et al. 2008).

Capitalism versus the Peasant

The World Bank's promotion of corporate agriculture as the solution to Africa's food production problems, after the devastation of structural adjustment, is a strong indication that, whether its designers were conscious of it or not, structural adjustment's main function was to serve as the cutting edge of a broader and longer-term process: the thoroughgoing capitalist transformation of the countryside.

That the dynamics of capitalist transformation lie at the heart of the food crisis is essentially what Oxford economist Paul Collier (2008) contends in his orthodox account of the causes and dynamics of the food price crisis. A large part of the blame for the crisis stems from the failure to diffuse what he calls the "Brazilian model" of commercial farming in Africa and the persistence of peasant agriculture globally. Despite what he knows to be the negative environmental impacts associated with the Brazilian model, Collier uses the term to underline that capitalist industrial agriculture, introduced in the United States and now being perfected by Brazilian enterprises for developing country contexts, is the only viable future if one is talking about global food production keeping up with global population growth. The peasantry is in the way of this necessary transformation. Peasants, he says, are not entrepreneurs or innovators, being too concerned with their own food security. They would rather have jobs than be entrepreneurs, for which only a few people are fit. The most capable of fitting the role of innovative entrepreneurs are commercial farming operations:

Reluctant peasants are right: their mode of production is ill suited to modern agricultural production, in which scale is helpful. In modern agriculture, technology is fast-evolving, investment is lumpy, the private provision of transportation infrastructure is necessary to counter the lack of its public provision, consumer food chains are fast-changing and best met by integrated marketing chains, and regulatory standards are rising towards the holy grail of traceability of produce back to its source. (Collier 2008: 71)

In his dismissal of peasant agriculture, Collier is joined by many, including scholars otherwise sympathetic to the plight of the peasantry and rural workers such as Henry Bernstein, who claims that advocacy of the peasant way "largely ignores issues of feeding the world's population, which has grown so greatly almost everywhere in the modern epoch, in significant part because of the revolution in productivity achieved by the development of capitalism" (2009: 255). Indeed, some progressives have already written off the peasantry, with the eminent Eric Hobsbawm declaring in his influential book, *The Age of Extremes*, that the "the death of the peasantry" was "the most dramatic and far-reaching social change of the second half of [the 20th] century," one that cut "us off forever from the world of the past" (1994: 298).

The Brazilian agro-enterprise that Collier touts as the solution to the food crisis is a key element in a global agrifood system where the export-oriented production of meat and grain is dominated by large industrial farms with global supply chains like those run by the Thai multinational Charoen Pokphand and where technology is continually upgraded by advances in genetic engineering from firms like Monsanto. The global integration of production is accompanied by the elimination of tariff and non-tariff barriers to facilitate the creation of a global agricultural supermarket of elite and middle-class consumers serviced by grain-trading corporations like Cargill and Archer Daniels Midland and transnational food retailers like the British-owned Tesco and French-owned Carrefour. These processes of integration and liberalization are governed by a multilateral superstructure, the centrepiece of which is the WTO.

According to Harriet Friedmann, the "dominant tendency" in the contemporary agrifood system,

is toward *distance* and *durability*, the suppression of particularities of time and place in both agriculture and diets. More rapidly and deeply than before, transnational agrifood capitals disconnect production from consumption and relink them through buying and selling. They have created an integrated productive sector of the world economy, and peoples of the Third World have been incorporated or marginalized — often simultaneously — as consumers and producers. (1994: 272)

Indeed, there is little room for the hundreds of millions of rural and urban poor in this integrated global market. They are confined to giant suburban *favelas*, where they have to contend with food prices that are often much higher than the supermarket prices, or to rural reservations, where they are trapped in marginal agricultural activities and are increasingly vulnerable to hunger.

These developments constitute not simply the erosion of national food self-sufficiency or food security but what some students of agricultural trends call "de-peasantization" — the phasing out of a mode of production to make the countryside a more congenial site for intensive capital accumulation (Bryceson 2000). This transformation has been a traumatic one for hundreds of millions of people, since peasant production is not simply an economic activity. It is an ancient way of life, a culture, which is one reason displaced or marginalized peasants in India have taken to committing suicide. In the state of Andhra Pradesh, farmer suicides rose from 233 in 1998 to 2,600 in 2002; in Maharashtra, suicides more than tripled, from 1,083 in 1995 to 3,926 in 2005 (Patnaik 2004). One estimate is that some 150,000 Indian farmers have taken their lives over the last few years (*Hindu* 2007). Global justice activist Vandana Shiva (2004) explains why: "Under globalization, the farmer is losing her/his social, cultural, economic identity as a producer. A farmer is now a 'consumer' of costly seeds and costly chemicals sold by powerful global corporations through powerful landlords and money lenders locally."

Resistance

Yet peasants have refused to go gently into that good night to which Collier and Hobsbawm — not to say Marx — would consign them. Indeed, one year before Hobsbawm's book was published, in 1993, La Vía Campesina was founded, and this federation of peasants and small farmers has become an influential actor on the agriculture and trade scene globally. The spirit of internationalism and active identification of one's class interests with the universal interest of society, which was once a prominent feature of workers' movements, is now on display in the international peasant movement. La Vía Campesina and its allies hotly dispute the inevitability of the hegemony of capitalist industrial agriculture, claiming that peasants and small farmers continue to be the backbone of global food production, constituting over a third of the world's population and two thirds of the world's food producers (Wayne Roberts, cited in McMichael 2008). Indeed, agroecologist Miguel Altieri reports:

> Millions of small farmers in the Global South still produce the majority of staple crops needed to feed the planet's rural and urban populations. In Latin America, about 17 million peasant production units occupying close to 60.5 million hectares, or 34.5% of the total cultivated land with average farm sizes of about 1.8 hectares, produce 51% of the maize, 77% of the beans, and 61% of the potatoes for domestic consumption. Africa has approximately 33 million small farms, representing 80 percent of all farms in the region. Despite the

fact that Africa now imports huge amounts of cereals, the majority of African farmers (many of them women) who are smallholders with farms below 2 hectares, produce a significant amount of basic food crops with virtually no or little use of fertilizers and improved seed. In Asia, the majority of more than 200 million rice farmers… make up the bulk of the rice produced by Asian small farmers. (Altieri 2008)

The food price crisis, proponents of peasant and smallholder agriculture claim, is not due to the failure of peasant agriculture but to that of corporate agriculture. They say that, despite the claims of its representatives that corporate agriculture is best at feeding the world, the creation of global production chains and global supermarkets, driven by the search for monopoly profits, has been accompanied by greater hunger, worse food and greater agriculture-related environmental destabilization all around than at any other time in history (La Vía Campesina 2008). Moreover, they assert that the superiority in terms of production of industrial capitalist agriculture is not sustained empirically. Altieri and Clara Nicholls, for instance, refute the conventional wisdom that small farms are backward and unproductive:

> Research shows that small farms are much more productive than large farms if total output is considered rather than yield from a single crop. Small integrated farming systems that produce grains, fruits, vegetables, fodder, and animal products outproduce yield per unit of single crops such as corn (monocultures) on large-scale farms. (Altieri and Nicholls 2008: 474)

When one factors in the ecological destabilization that has accompanied the generalization of capitalist industrial agriculture, the balance of costs and benefits lurches sharply towards the negative. For instance, in the United States, notes Daniel Imhoff:

> The average food item journeys some 1300 miles before becoming part of a meal. Fruits and vegetable are refrigerated, waxed, colored, irradiated, fumigated, packaged, and shipped. None of these processes enhances food quality but merely enables distribution over great distances and helps increase shelf life. (1996: 425–26)

Industrial agriculture has created the absurd situation whereby "between production, processing, distribution, and preparation, 10 calories of energy are required to create just one calorie of food energy" (Imhoff 1996: 426). Conversely, its ability to combine productivity and ecological sustainability constitutes a key dimension of the superiority of peasant and small-scale agriculture over industrial agriculture.

Contrary to assertions that peasant and small-farm agriculture is hostile to technological innovation, partisans of small-scale and peasant-based farming assert that technology is "path dependent," that is, its development is conditioned by the mode of production in which it is embedded, so that technological innovation under peasant and small-scale farming would take different paths than innovation under capitalist industrial agriculture. The development of small intensive integrated multicrop farming systems, with their proven superiority in production and productivity to monoculture while keeping farmers on the land and promoting environmentally sound practices, is one example of this interaction of social organization and technological direction.

But partisans of the peasantry have not only engaged in a defence of peasant and smallholder agriculture. La Vía Campesina and its allies have formulated a desperately needed alternative to industrial capitalist agriculture, one that looks to the future rather than to the past: the paradigm of food sovereignty.

The Conjuncture

To be fully understood, the global food price crisis of the last few years, which is essentially a crisis of production, must be seen as a critical juncture in the centuries-long process of displacement of peasant agriculture by capitalist agriculture. Despite its dominance, capitalist agriculture never really managed to eliminate peasant and family farm-based agriculture, which has survived until now and continues to provide a substantial part of the food supply for national populations, particularly in the South.

Yet, even as capitalism seems poised to fully subjugate peasant agriculture, its dysfunctional character is being fully revealed. For it has not only condemned millions to marginalization but also imposed severe ecological costs, especially in the form of severe dependency on fossil fuels at all stages of its production process, from the manufacture of fertilizers, to the running of agricultural machinery, to the transportation of its products. Indeed, even before the food price crisis and the larger global economic crisis of which it was a part, the legitimacy of capitalist industrial agriculture was eroding, and resistance to it was rising, not only from the peasants and small farmers it was displacing but from consumers, environmentalists, health professionals and many others who were disconcerted by the mixture of corporate greed, social insensitivity and reckless science that increasingly marked its advance.

Now, with the collapse of the global economy, the integration of production and markets that has sustained the spread of industrial agriculture is going into reverse. "Deglobalization" is in progress "on almost every front," says the *Economist*, adopting a word that Walden Bello (2002) coined nearly a decade ago. The magazine, probably the most vociferous cheerleader of globalization, warns that the process depends on the belief of capitalist enterprises "in the efficiency of global supply chains. But like any chain, these are only as strong as their weakest link. A danger point will come if firms decide that this way of organizing production has had its day" (2009: 59). The next few years — nay, months, given the speed with which the global economy is plunging into depression — will provide the answer.

As the capitalist mode of production enters its worst crisis since the 1930s, peasants and small farmers increasingly present a vision of autonomy, diversity and cooperation that may just be the key elements of a necessary social and economic reorganization. As environmental crises multiply, the social dysfunctions of urban industrial life pile up and globalization drags the world to a global depression, the "peasant's path" has increasing relevance to broad numbers of people beyond the countryside.

Indeed, not only in the South but also in the North, there are increasing numbers who seek to escape the dependency on capitalist corporations by reproducing the peasant condition, where one works with nature from a limited resource base to create a condition of relative autonomy from the forces of capital and the market. The emergence of urban agriculture, the creation of networks linking consumers to farmers within a given region, the rise of new militant movements for land — all this according to Jan Douwe van der Ploeg may point to a movement of "repeasantization," which has been created by the negative dynamics of global capitalism and empire and seeks to reverse them. Under these conditions, "the peasant principle, with its focus on the construction of an autonomous and self governed resource base, clearly specifies the way forward" (2008: 276). Along the way, articulating food sovereignty with other broader approaches that stress the principles of small size, subsidiarity, diversity, equality and democracy while at the same time going beyond production of food to also consider what implications food sovereignty has for workers, industry, services and other sectors of the economy will be key.

Note

This chapter is a revised version of "Food Wars," which appeared in *Monthly Review*, July 2009.

References

AFP. 2009. "Global Trends Driving 'Land Grab' in Poor Nations: Activists." Available at <google. com/hostednews/afp/article/ALeqM5iAAAFho9FSMtNoh1BfnqWlgFT5LQ>.

Altieri, M. 2008. "Small Farms as a Planetary Ecological Asset: Five Key Reasons Why We Should Support the Revitalization of Small Farms in the Global South." Food First. Available at <foodfirst.org/en/node/2115>.

Altieri, M., and C. Nicholls. 2008. "Scaling up Agroecological Approaches for Food Sovereignty in Latin America." *Development* 51, 4.

APEC (Asia-Pacific Economic Cooperation). 2008. APEC Biofuels Website. Available at <biofuels.apec.org/me_ur. ted_states.html>.

Bartra, A. 2004. "Rebellious Cornfield: Towards Food and Labor Self-Sufficiency." In G. Otero (ed.), *Mexico in Transition*. London: Zed Books.

Bello, W. 2002. *Deglobalization: Ideas for a New World Economy*. London: Zed Press.

Bernstein, H. 2009. "Agrarian Questions from Transition to Globalization." In A.H. Akram-Lodhi and C. Kay (eds.), *Peasants and Globalization*. New York: Routledge.

Bryceson, D. 2000. "Disappearing Peasantries? Rural Labor Redundancy in the Neo-Liberal

Era and Beyond." In C.K. Bryceson and J. Mooij (eds.), *Disappearing Peasantries? Rural Labor in Africa, Asia, and Latin America.* London: Intermediate Technology Publications.

Chakrabortty, A. 2008. "Secret Report: Biofuels Caused Food Crisis." *Guardian,* July 3.

Collier, P. 2008. "The Politics of Hunger: How Illusion and Greed Fan the Food Crisis." *Foreign Affairs* 87, 6.

Economist. 2009. "Turning Their Backs on the World." February 21–27.

FAO (Food and Agriculture Organization). 2008. "Briefing Paper: Hunger on the Rise." Rome.

Friedmann, H. 1994. "Distance and Durability: Shaky Foundations of the World Food Economy." In P. McMichael (ed.), *The Global Restructuring of Agro-Food Systems.* Ithaca, NY: Cornell University Press.

Havnevik, K., D. Bryceson, L.E. Birgegard, P. Matondi and A. Beyene. 2008. "African Agriculture and the World Bank." *Pambazuka News,* March 11.

Hindu. 2007. November 12. Available at <hindu.com/2007/11/12/stories/2007111257790100.htm>.

Hobsbawm, E. 1994. *The Age of Extremes: The Short Twentieth Century, 1914–1991.* London: Abacus.

Imhoff, D. 1996. "Community Supported Agriculture." In J. Mander and E. Goldsmith (eds.), *The Case Against the Global Economy.* San Francisco: Sierra Club.

La Vía Campesina. 2008. "An Answer to the Global Food Crisis: Peasants and Small Farmers Can Feed the World!" Available at <viacampesina.org>.

Lindsay, R. 2008. "Inside Haiti's Food Riots." *Al-Jazeera,* April 16.

McMichael, P. 2008. *Food Sovereignty in Movement: The Challenge to Neo-liberal Globalization.* Draft. Cornell University.

Mitchell, D. 2008. "A Note on Rising Food Prices." World Bank. Available at <go.worldbank.org/31PG0020G0>.

Patnaik, U. 2004. "External Trade, Domestic Employment, and Food Security: Recent Outcomes of Trade Liberalization and Neo-Liberal Economic Reforms in India." Paper presented at the International Workshop on Policies against Hunger III, Berlin, October 20–22.

Shiva, V. 2004. "The Suicide Economy." Znet Website. Available at <countercurrents.org/glo-shiva050404.htm>.

Spencer, R. 2008. "South Korean Company Takes Over Part of Madagascar to Grow Biofuels." *Telegraph* November 20. Available at <telegraph.co.uk/earth/agriculture/3487668/South-Korean-company-takes-over-part-of-Madagascar-to-grow-biofuels.html>.

United Nations. 2009. *World Economic Situation and Prospects.* New York.

van der Ploeg, J.D. 2008. *The New Peasantries.* London: Earthscan.

Wahl, P. 2009. *Food Speculation: The Main Factor of the Price Bubble in 2008.* Berlin: WEED.

World Bank. 2008a. "Accelerating Inclusive Growth and Deepening Fiscal Stability: Draft Report for the Philippines Development Forum." Manila.

_____. 2008b. *World Bank Development Report 2008: Agriculture for Development.* Washington, DC.

6

Agrofuels & Food Sovereignty
Another Agrarian Transition

Eric Holt-Giménez & Annie Shattuck

In 2008, 5 percent of the world's corn crop was converted to liquid vehicle fuels. Called a "crime against humanity" by the UN Special Rapporteur on the Right to Food, the agrofuels boom detonated the 2007–08 food price spike that sent an additional 100 million people into hunger. Since production began to rapidly increase in 2005, agrofuels have been surrounded by controversy. While the debate around crop-based fuels often engages the term "biofuels," we are concerned with agrofuels — mostly ethanol and biodiesel produced on a large agro-industrial scale from crops such as corn, sugarcane and soy, largely produced for the global market. Biofuels on the other hand refers to small-scale, non-industrial produced fuels, often made in owner-operated facilities for local use.

Thus far, most debates on agrofuels revolve around energy balance, economic and environmental cost-benefits and food and energy security. Few analysts have focused on the ways in which the industrial development of agrofuels is fundamentally transforming the world's food and fuel systems. New corporate partnerships, increased market power and concentration of land ownership in the Global South are changing the dynamics of food and fuel systems worldwide. The convergence of these forces has far reaching implications that may not only lead to irreversible environmental impacts but will exacerbate problems of land tenure, food security, migration and poverty worldwide.

While industrial, internationally traded agrofuels have only appeared in the last five to ten years, the boom follows on two hundred years of industrialization of the food system. The agrofuels boom is the latest development in a relationship between agriculture and industry that began with the Industrial Revolution and led to the global industrialization of agriculture itself. The agrofuels boom has been engineered by Northern countries whose renewable fuel mandates and massive agrofuels subsidies are leading to extensive corporate enclosures throughout the developing world (Holt-Giménez 2008).

At the same time, transnational social movements are advocating for food sovereignty, people's right to abundant, healthy, culturally appropriate food and the democratization of the food system in favour of the poor and underserved. The international peasant movement La Vía Campesina and its allies have made some very real progress towards creating greater equity and democratic control in the food system. That progress, however, is under attack by the agrofuels boom.

This chapter examines the myths that package agrofuels as an environmentally sustainable, renewable energy and rural development policy. We contend that these myths hide the agrarian transformations lurking behind the agrofuels boom and discuss the economic drivers, corporate consolidation and role of international finance. We conclude with a discussion of the territorial restructuring taking place and outline the challenges to building sovereign food systems in the face of an expanding agrofuels economy.

Agrofuels Myths

"Biofuels" invoke an image of renewable abundance that allows industry, politicians, the World Bank, the UN and even the Intergovernmental Panel on Climate Change to present fuel from corn, sugarcane, palm, soy and other crops as allowing a smooth transition from peak oil to a renewable fuel economy. Myths of abundance divert attention away from powerful economic interests that benefit from what in fact is part of an agrarian transition and avoid discussion of the growing price that citizens of the Global South and underserved communities in the Global North are beginning to pay to maintain consumptive oil-based lifestyles in the North. Myths around agrofuels production obscure the profound consequences of the further industrial transformation of our food and fuel systems.

Myth #1: Agrofuels are clean and green

Because photosynthesis removes greenhouse gases from the atmosphere and can reduce fossil fuel consumption, we are told that fuel crops are green (Inhofe 2007). But when the full life cycle of agrofuels is considered — from clearing land for plantations, to planting, processing, shipping and automotive consumption — the moderate emission savings are undone by far greater emissions from deforestation, burning, peat drainage, cultivation and soil carbon losses. Every ton of palm oil produced results in thirty-three tons of carbon dioxide emissions — ten times more than petroleum (Monbiot 2007). Tropical forests cleared for sugarcane ethanol emit 50 percent more greenhouse gases than the production and use of the same amount of gasoline (Tillman and Hill 2007). Studies demonstrate that when full life-cycle and land-use changes are considered, agrofuels may emit more greenhouse gases than fossil fuels (Crutzen et al. 2007; Searchinger et al. 2008; Jacobsen 2009; Howarth et al. 2009). Industrial agrofuels require large applications of petroleum-based fertilizers, whose global use has more than doubled the biologically available nitrogen in the world, contributing heavily to the emission of nitrous oxide, a greenhouse gas three hundred times more potent than CO^2 (Crutzen et al. 2007).

There are other environmental problems as well. To produce a litre of ethanol takes three to five litres of irrigation water and produces up to thirteen litres of waste water. It takes the energy equivalent of 113 litres of natural gas to treat this waste, increasing the likelihood that it will simply be released into the environment to pollute streams, rivers and groundwater (Aslow 2007). Intensive cultivation of fuel crops also leads to high rates of erosion, particularly in soy production — from 6.5 tons per hectare in the U.S. to up to 12 tons per hectare in Brazil and Argentina (Altieri and Bravo 2007).

Myth #2: Agrofuels will not result in deforestation

Proponents of agrofuels argue that fuel crops planted on ecologically degraded lands will improve, rather than destroy, the environment. Perhaps the government of Brazil had this in mind when it reclassified some 200 million hectares of dry-tropical forests, grassland and marshes as "degraded" and apt for cultivation (Moreno 2006). In reality, these are the biodiverse ecosystems of the Mata Atlantica, the Cerrado and the Pantanal, occupied by indigenous people, subsistence farmers and extensive cattle ranches. If historical patterns continue, the introduction of agrofuel plantations will simply push these communities to the "agricultural frontier" of the Amazon, where deforestation will intensify (Holt-Giménez and Patel 2009). NASA, the U.S. National Aeronautics and Space Administration, has positively correlated the market price of soybeans, a main biodiesel feedstock, with satellite data on the rate of destruction of the Amazon rainforest — nearly 325,000 hectares a year (Morton et al. 2006). Called the "diesel of deforestation," oil-palm plantations for biodiesel are the primary cause of forest loss in Indonesia, a country with one of the highest deforestation rates in the world. By 2020, Indonesia's oil-palm plantations will triple in size to 16.5 million hectares — an area the size of England and Wales combined — resulting in a loss of 98 percent of forest cover (Aslow 2007).

Myth #3: Agrofuels will bring rural development

In the tropics, 100 hectares dedicated to family farming generates thirty-five jobs, while oil palm and sugarcane provide ten jobs, eucalyptus two and soybeans just one-half job per 100 hectares, all poorly paid (FBOMS 2006). Until this boom, crop-based fuels primarily supplied local markets, and even in the U.S., most ethanol plants were small and farmer-owned. Big Oil, Big Grain and Big Genetic Engineering are rapidly consolidating control over the entire agrofuel value chain. These corporations enjoy immense market power. Bunge, Cargill and Archer Daniels Midland (ADM) together control 90 percent of the global grain trade (O'Driscoll 2005). In 2008, 80 percent of corn, 92 percent of soy and 86 percent of cotton in the U.S. were under patent from major seed companies (USDA 2008). This market power, which is highly concentrated in most sectors of the agricultural economy, allows corporations to extract profits from the most lucrative and low-risk segments of the value chain — inputs, processing and distributing. In the long run, profits will accumulate to the processing, seed and agro-chemical monopolies, not to farmers (Dufey 2007). Smallholders in the Global South are increasingly being forced off the land. For example, hundreds of thousands of smallholder farmers and indigenous people have already been displaced by soybean plantations in the "Republic of Soy," a 50+ million hectare area covering Southern Brazil, Northern Argentina, Paraguay and eastern Bolivia (Bravo 2006).

Myth #4: Agrofuels will not cause hunger

Hunger, said Nobel prize-winning economist Amartya Sen, results not from scarcity but poverty (Sen 1981). Food production has been outstripping population growth for decades. Nonetheless, global hunger continues to grow (Holt-Giménez and Patel 2009), and because they are poor, nearly a billion people continue to go hungry (FAO 2008b). In 1996, at the World Food Summit in Rome, world leaders promised to halve the proportion of hungry people living in extreme poverty by 2015. Little progress has been made. The world's poorest people already spend 50 to 80 percent of their total household income on food (FAO 2008c). They suffer when high fuel prices push up food prices. Now, because food and fuel crops are competing for land and resources, high food prices may actually push up fuel prices. Both increase the price of land and water. This perverse, inflationary spiral puts food and productive resources out of reach for the poor. While reports varied, the highest claimed agrofuels were responsible for 75 percent of 2008's food price inflation (Senauer 2007); others pointed to the increased volatility and speculative activity spurred by the agrofuels boom (Holt-Giménez 2008). Regardless, it is clear that agrofuels played a significant role in the food price spike (Holt-Giménez 2008). Volatile and rising food prices are particularly dangerous; caloric consumption typically declines as price rises by a ratio of 1:2. With every one percent rise in the cost of food, 16 million people are made food insecure. If current trends continue, some 1.2 billion people could be chronically hungry by 2025 — 600 million more than previously predicted (Runge and Senaur 2007). World food aid will not likely come to the rescue because surpluses will go into our gas tanks. Last year, world food aid reached its lowest level since 1961 (Holt-Giménez 2008), forcing the World Food Program to cut back on rations in Darfur and other conflict zones.

Myth #5: Better "second-generation" agrofuels are on the way

Proponents of agrofuels argue that present agrofuels made from food crops will soon be replaced with environmentally friendly crops like fast-growing trees and switchgrass (Inhofe 2007). This myth, wryly referred to as the "bait and switch-grass" shell game, makes crop-based fuels socially acceptable. The agrofuel transition transforms land use on a massive scale, pitting food production against fuel production for land, water and favourable policy conditions. The issue of which crops are converted to fuel is irrelevant. Wild plants cultivated as fuel crops won't have a smaller environmental footprint. They are likely to rapidly migrate from hedgerows and woodlots onto arable lands to be intensively cultivated like any other industrial crop, with all the associated environmental externalities. Further, major discoveries in plant physiology are still required that permit the economically efficient breakdown of cellulose, hemi-cellulose and lignin. Industry is either betting on miracles or counting on taxpayer bail-outs. A University of Iowa study found that without the subsidies offered by the U.S.'s 2007 *Energy Act*, cellulosic ethanol would never get off the ground (Baker et al. 2008). Even with the Renewable Fuels Standard, the authors claim that the U.S. will fall drastically short of targets, producing around 4.5 billion of the mandated 21 billion gallons of "advanced biofuels" per year, and then only if the government triples the already enormous per gallon subsides to the ethanol industry (Baker et al. 2008). Asking taxpayers to launch cellulosic is illogical at best, especially considering that a 3–4 percent increase in fuel economy would save the same amount of fuel the Iowa study predicts will be displaced by cellulosic even with massive subsidies (Baker et al. 2008).

The Agrofuels Boom

Cloaked in myths, the changes in global food systems due to the agrofuels boom become invisible. In reality, agrofuels are not an energy and climate change solution but an opportunity for further agri-business expansion — and the massive displacement of rural populations.

Over the last thirty years, agricultural production has outpaced global purchasing power, leading to classic capitalist overproduction and a steady decline in agribusiness's profit margins. In the past, agrifood corporations have responded to this falling rate of profit by increasing productivity with technological improvements (e.g., the Green Revolution), by adding value to raw commodities by transforming them (e.g., corn into beef), by creating and capturing proprietary benefits (e.g., genetically engineered seeds) or by subsidizing, consolidating and "vertically integrating" their operations from production to market to capture more of the food value chain. Agrofuels does all of this in one industrial operation. They are a one-stop-shop for solving agri-business's overproduction problem. The transformation of food into fuel inflates the value of overproduced commodities like corn and sugarcane in both food and fuel markets, opens up new market space for those commodities and creates more processing steps, which allow corporate players to both add and capture more value.

Prices and Profits

Agrofuels ignited the explosion in commodity prices that unleashed the 2007–08 food crisis. The FAO (2008a) food price index rose 60 percent during that period. In December 2007 the *Economist*'s food price index was at the highest it had been since its creation in 1845 (*Economist* 2007a). By March 2008, wheat prices were up 137 percent from the year before, soy was up 87 percent, rice 74 percent and maize 31 percent (Holt-Giménez 2008). While global commodity prices have since come down (though not to pre-crisis levels), retail food prices in the developing world have not come down as far, as fast or in many cases at all (FAO 2009).

Global grain traders reaped windfall profits. Corporations like Cargill and

ADM. which together control 75 percent of the global grain trade (Vorley 2003), thrived on the crisis. The benefits from market volatility in grains to these monopolies were clearly demonstrated in their record earnings. In a time of severe economic downturn, when most companies suffered enormous losses, Cargill's earnings increased 62 percent for the quarter ending August 31, 2008. over the same quarter of 2007 (Black 2008). Net income at Bunge, one of the top three global grain traders, increased 471 percent in the first half of 2008 (Ugarte and Murphy 2008). After the boom in grain prices abated, a Cargill representative actually bemoaned the lack of volatility in commodities markets in the first quarter of 2009. The relatively stable market made short selling grain futures impossible and cut into the company's exploding profits (Weitzman 2009).

ADM, the largest U.S. (and global) grain processor, gets 25 percent of its operating profit from agrofuels, including both ethanol and biodiesel (Scully 2007). In anticipation of the passing of the 2007 U.S. Energy Bill, ADM's stock surged nearly 20 percent from August to mid December (Philpott 2007). The company announced that it was "optimistic about the expanded role [agrofuels] will play in improving energy security, strengthening rural economies and helping to improve our environment" (ADM 2007).

Concentrating Market Power

In the United States, agrofuels began as a locally owned, homegrown industry largely controlled by farmers' cooperatives in the Midwest. In the late 1990s, the majority of new ethanol plants were farmer-owned, and by 2002 farmer-owned plants were collectively out-producing industry giant ADM (Morris 2005). But when a glut of international investment capital was combined with a package of government mandates, subsides and high oil prices, the "green gold rush" was on. Massive investment poured into the sector, quickly shifting the ownership structure towards large corporate monopolies.

From 1998 to 2000, the number of agrofuel producing facilities in the United States grew from thirty-eight to seventy-five (FTC 2005). These seventy-five producers operated the ninety existing agrofuel production plants, many of them farmer-owned (FTC 2005). The agrofuel industry began to consolidate in 2004, when corporations such as ADM started purchasing farmer-owned cooperatives and investing in new production capacity (Morris 2005). By 2007, five corporations controlled roughly 47 percent of all ethanol production in the U.S. (Hassan 2007). ADM and POET, the two largest corporate ethanol producers, controlled 33.7 percent of all ethanol production, and the top ten producers together controlled an estimated 70 percent (Hassan 2007).

Since 2007 and the introduction of the U.S. renewable fuels targets, the industry has moved even further away from farmer and locally owned plants. According to the Renewable Fuels Association (RFA), the ethanol industry's lobby group, out of a total of two hundred operational ethanol-processing plants in the U.S., forty were locally owned as of January 2010 — accounting for a scant 16 percent of the nation's refining capacity (RFA 2010). This is vastly different from where the industry began. As recently as May 2007, farmer-owned plants were responsible for 34 percent of overall production (Hassan 2007). Of twenty ethanol plants currently under construction, none are farmer or locally owned (RFA 2010).

Because of the economies of scale at ADM's plants, its dominance in the global grain trade and its disproportionate access to government subsidies, ADM is emerging as a hegemonic player in the U.S. When other ethanol companies were struggling with shrinking margins due to high corn prices, ADM has strengthened its market share and its profits (Birger 2008).

Concentration of ownership in global agrofuels production is proceeding rapidly as well. U.S.-based Cargill is now the largest shipper of both raw sugar and soybeans from Brazil — the former for ethanol feedstock, the latter for either feed or biodiesel. Cargill also has the largest processing capacity for oil seeds in Paraguay (GRAIN 2007a). In 2005, Cargill became the majority shareholder of two oil-palm plantations in Indonesia, on the islands of Sumatra and Borneo, and three more in Papua New Guinea. (For more examples of Cargill's acquisitions see GRAIN 2007b).

From 2004 to 2007, venture capital investment in agrofuels increased by nearly 700 percent (Reeves 2007). Private investment in agrofuels is also pouring into public research institutions, setting their research agenda not only for agrofuels but also for public scholarship in general (Altieri and Holt-Gimenez 2007). New corporate partnerships are also being formed between agri-businesses, biotechnology companies, oil companies and car manufacturers: Royal Dutch Shell with Cargill, Syngenta and Goldman-Sachs; British Petroleum (BP) with DuPont and Toyota, as well as with Monsanto and Mendel Biotechnology; ADM with both Monsanto and Conoco-Phillips; and DuPont with BP and Weyerhauser (ETC 2007).

With the global economy in tatters, agrofuels and the higher values for agricultural surplus they create offer an attractive investment for international capital. Leading venture capital firms in the U.S., such as Soros Fund Management, Warburg-Pincus, Kholsa Ventures and Goldman Sachs, have invested heavily in the new industry. International investment in agrofuels presents a risk, however, with resistance to agrofuels development mounting from both civil society and peasants and small farmers' groups. International financial institutions (IFIs) have stepped in to mitigate some of this risk. Institutions like the Inter-American Development Bank and the World Bank's private-sector lending arm, the International Finance Corporation (IFC), are using their nominal safeguard procedures to politically legitimize investment in agrofuels (Jonasse 2009). For example, the Inter-American Development Bank is providing direct loans to the second largest ethanol conglomerate in Brazil. Santelisa is a huge conglomerate that includes investments from all over the globe, including Goldman Sachs and the Carlyle Group. It is also involved with Cargill, BP and a whole host of Brazilian agro-industry leaders. The

IFC is providing direct loans to the largest Brazilian sugar conglomerate, Número Um (Jonasse 2009).

The Dutch financial group Rabobank cited the IFC certification as the reason it felt safe to invest in Brazilian ethanol: "Rabobank's reasoning was that if IFC approves this project and they classify it only as a class B, low risk project, we can safely invest [an additional] $230 million… in this corporation" (Lilley 2004). The myths around agrofuels have made it possible for IFIs to rate investments in highly destructive projects favourably. An example is the IFC's 2004 loans to Ammagi Soy for a project that was found to have destroyed large swaths of the Amazon Rainforest (Lilley 2004), securing the territory for further transnational investments.

Territorial Restructuring

The agrofuels boom must be understood in the context of the fundamental changes capitalism creates in both physical places and political and economic spaces (Harvey 2003). The analytical concept of territorial restructuring — a reshaping of both places and spaces at the national, international and regional levels (Holt-Gimenez 2007) — can help explain the changes occurring under the agrofuels transition and what they mean apart from the rosy discourse of "alternative energy." International financial institutions, corporate agri-business, large landowners, governments and market forces are all engaged in redrawing the lines of power and ownership — restructuring places and spaces not through some broad consensus but through the interaction between the interests and activities of the different actors. While IFIs restructure political and economic spaces by blazing the trail for international agrofuels investment, physical territory is being restructured by agri-business and biotech firms on the ground.

The agrofuels industry has caused a massive concentration in landholdings in the Global South. The agrofuels boom in Guatemala has led to "considerable loss in the amount of land available for food cultivation" and resulted in wide-spread land evictions, concentration of land ownership and violations of human rights (Hurtado 2009). In Colombia, according to one report, "93 percent of the land under palm cultivation… is located in the collective territorial zone of black communities," while nearly all traditional villages have been cleared and are being resettled with former paramilitaries and outsiders (Zimbalist 2007). In the Eastern Cape of South Africa, 500,000 hectares of communal farmland have been fenced and planted to canola for biodiesel. Local farmers have been forced to relinquish their traditional use of these areas for food gardens and grazing lands while Monsanto collects heavy subsidies for providing its chemicals and seeds "on the farmers' behalf" (African Centre for Biosafety 2008). From Tanzania to Sudan, massive land grabs in Africa for biofuels and food production have farmers and activist groups warning of a new era of colonialism in Africa (GRAIN 2008).

In another example, Aracruz Cellulose, a leading supplier of eucalyptus paper pulp and one of the new players in cellulosic ethanol, evicted 8,500 indigenous families from their land in the Brazilian state of Espirito Santo, converting 11 thousand hectares to "green desert" (Meirelles 2005). The plantations have dried up several rivers and streams, seriously threatening the water supply to small farmers (FOEI 2008). If the technology to commercialize cellulosic ethanol from wood products becomes widely available, as companies like Aracruz hope, more peasants and small farmers will be displaced into the agricultural frontier or worse, to urban slums, by the march of fuel crops into the Brazilian landscape.

Restructuring the Genetic Commons

The agrofuels transition, unlike previous incarnations of the agrarian transition/industrial revolution, will also enclose vast genetic resources in the private sector and create new economic space under the dominion of agri-business. Biotechnology companies are aggressively using the agrofuels boom to extend intellectual property rights over a greater percentage of the world's agricultural genetic resources and consolidate more market power by horizontally integrating seed, chemical packages and processing of agrofuels. Both Monsanto and Syngenta have recently come out with genetically modified varieties specifically for processing corn into ethanol, while development of "second generation" agrofuels is proceeding under the direction of large biotechnology firms.

In 2006, Monsanto and agri-business giant Cargill jointly launched a new corporation, Renessen, with an initial investment of $450 million. Renessen is the sole provider of the first commercially available genetically engineered energy crop, Mavera high-value corn. Mavera corn is stacked with genes coded for increased oil content and production of the amino acid lysine — an essential nutrient missing in cattle feedlot diets, along with Monsanto's standard Bt pesticide and herbicide tolerant traits. Farmers must sell their crop of Mavera corn to a Renessen-owned processing plant in order to recover the "higher value" of the crop, for which they paid a premium on the seed. Cargill's agricultural processing division created a plant for Renessen that only processes Renessen-brand corn. Due to the engineered production of lysine in the corn, Renessen can sell the waste stream from its ethanol processing facilities as a value-added cattle feed (Shattuck 2008). Renessen is an attempt by Monsanto and Cargill to achieve perfect vertical integration. Renessen sets the price of seed, Monsanto sells the seed and agro-chemical inputs, Renessen sets the price at which to buy back the finished crop, Renessen sells the fuel and feed, and farmers are left to absorb the risk. This system robs small farmers of choices and market power, while ensuring maximum monopoly profits for Renessen/ Monsanto/Cargill (Shattuck 2008).

Biotech's foray into energy does not stop with corn. The industry promises a second generation of cellulose-based energy crops that can grow on the marginal land passed over by previous Green Revolution technologies. Cellulosic energy crops can conceivably be produced from any plant material: corn stalks, trees, sugarcane biomass or grasses. Overcoming the key stumbling blocks to cellulosic energy — processing efficiency and yield — offers the industry unprecedented opportunity to extend their market power and enclose more genetic material – the building blocks of agricultural production — under private patent.

Plants like the genetically modified high biomass "energycane" (or sorghum) being developed by Ceres (a small biotech start-up with significant equity investment from Monsanto) are engineered to trade increased biomass yield for their ability to produce a food product. Trees like eucalyptus, poplar and radiata pine are being genetically engineered to produce less lignin — which will aid in pulping and future conversion to ethanol. Companies like Mendel Biotechnology, with heavy investments from Monsanto and British Petroleum, are engineering patented varieties of the weedy grass miscanthus, along with its own sorghum and energycane varieties.

Processing cellulose into sugars is the largest hurdle in making cellulosic ethanol practical. At its current stage, processing is vastly inefficient. The engineering of new enzymes and bacteria that break down cellulose is a multimillion dollar race. Corporate partnerships, and not competition, is the norm in this sector. Codexis, one of the leading developers of genetically engineered enzymes, is partnering with Syngenta and Shell Oil for its research and development. Some enzyme biotechnology firms like the Kholsa Ventures-funded company, Range Fuels, also own ethanol processing plants (Shattuck 2008). Patents on this technology will essentially put a stranglehold on the cellulosic ethanol market: whoever controls the most efficient catalysts will have a virtual monopoly on processing fuel.

The development of second-generation fuel crops represents another restructuring of places and spaces in our food systems. This development threatens to enclose more genetic material in private spheres. Inserting one or two novel traits effectively extends ownership over all the genetic material contained in an altered seed. While this has been ongoing since the mid 1990s, when biotech products first began to go commercial, the scale of the restructuring of ownership of genetic resources the agrofuels boom allows is without precedent. The agrofuels boom gives cause for trees, grasses, non-traditional crops, enzymes, bacteria and even novel life forms to come under patent (Shattuck 2008). Second generation ethanol also allows more vertical integration in the industry than any previous incarnation of biotechnology.

The Agrofuels Transition

The agrofuels boom closes a historical chapter in the relation between agriculture and industry that dates back to the Industrial Revolution. Then, peasant agriculture effectively subsidized industry with both cheap food and cheap labour. Later on, cheap oil and petroleum-based fertilizers opened up agriculture to industrial capital. Mechanization intensified production, keeping food prices low and industry booming. Half of the world's population was pushed out of the countryside and into the cities (*Economist* 2007b). The massive transfer of wealth from agriculture to industry, the industrialization of agriculture and the rural-urban shift are all part of the "Agrarian Transition," the lesser-known twin of the Industrial Revolution. The agrarian/industrial twins transformed most of the world's fuel and food systems and established non-renewable petroleum as the foundation of today's multi-trillion dollar agrifoods complex. Like the original agrarian transition, the present "agrofuels transition" encloses public and common land by burning and plowing up the remaining forests, swamps, savannas and prairies of the world. It drives the planet's remaining smallholders, family farmers and indigenous peoples to the cities and funnels rural resources to urban centres in the form of fuel, generating massive amounts of corporate wealth.

Unfortunately, the agrofuels transition suffers from a congenital flaw: its fraternal twin is dead. There is no new industrial revolution. No expanding industrial sector waits to receive displaced indigenous communities, smallholders and rural workers. There are no production breakthroughs poised to flood the world with cheap food. This time, fuel will not subsidize agriculture with cheap energy. On the contrary, fuel will compete with food for land, water and other resources. Agrofuels collapse the industrial link between food and fuel. Taken to its extreme, agrofuel will be used to grow agrofuel — a thermodynamic dead end. The inherent entropy of industrial agriculture was invisible as long as oil was abundant and fueled food production. Now, with agrofuels, fuel systems feed on food systems.

Food Sovereignty: From Extraction to Redistribution

Food sovereignty is anathema to agrofuels because it aims to restructure territories in ways that restore and redistribute rather than concentrate resources; it decentralizes and democratizes power in the food system rather than concentrates it within a handful of corporate monopolies. However, the massive scale of territorial restructuring underway in the agrofuels boom poses formidable challenges to food sovereignty movements.

Resisting the agrofuels transition requires understanding how the "logic of capital" and the "logic of territory" inherent in capitalism restructure both physical places and political and economic spaces in order to extract wealth (Harvey 2003). Territorial restructuring redraws the lines of power and ownership in land-based production in order to facilitate extraction (Holt-Giménez 2007). By promoting agrofuels, international finance institutions, governments and agribusiness reorganize social, political and economic conditions for the accelerated extraction of wealth in biomass-rich territories. The agrofuels industry is an easy vehicle for extractive territorial restructuring because it fits neatly into present fuel consumption patterns, is reinforced by simplistic environmental and energy myths and reinforces the dominant industrial model of capitalist agriculture.

However, the agrofuels transition is reflective of a new, perverse and self-destructive phase in agrarian capitalism. In terms of net energy output, food production and carbon capture, this phase offers increasingly diminishing returns to society. Agrofuels are a modern-day, capitalist reflection of what Geertz (1963) understood historically as "agricultural involution," in which the ever-increasing social and environmental costs of production render ever declining social and environmental returns — eventually destroying the productivity and social structures of the system. In this regard, agrofuels are a reflection of the deep flaws of capitalist agriculture that undermine not only society but also agriculture itself. For all its political and economic power, agrofuels are the Achilles' heel of capitalist agriculture.

Food sovereignty as a political concept can expose the dangerous contradictions of agrofuels and advance sustainable solutions to the food, fuel and climate crises. By advancing a countervailing process — one of increasing social and environmental benefits and a more equitable distribution of resources and political power in the food system — food sovereignty offers up a powerful social discourse to counter the agrofuels myths. It is also a potential bridge between agrarian movements, food activists, conservationists and environmentalists. As an analytical concept, food sovereignty can also help identify and strategically differentiate these potential allies from those sectors in the food and energy industries (e.g., some petroleum companies, the meat industry and supermarket chains) that oppose agrofuels but seek to concentrate their own corporate power over food systems.

To roll back the agrofuels transition, food sovereignty movements will not only need strong transnational and national alliances, they will need allies on the ground, countering agrofuels' extractive onslaught with strategies for sustainable and redistributive territorial restructuring. In the words of the World Social Forum, "another world is possible." Inverting the agrofuels transition will help make "another agrarian transition" possible. Given the stakes, it is not only possible but necessary.

References

African Centre for Biosafety. 2008. "Rural Communities Express Dismay: 'Land Grabs' Fuelled by Biofuel Strategy." Available at <biosafetyafrica.net/portal/images//rural-communitiesexpressdismay.pdf>.

Altieri, M., and E. Bravo. 2007. "The Ecological and Social Tragedy of Crop-Based Biofuel Production in the Americas." Institute for Food and Development Policy. Available at <foodfirst.org/en/node/1662>.

Altieri, M., and E. Holt-Giménez. 2007. "UC's Biotech Benefactors: The Power of Big Finance and Bad Ideas." *The Berkeley Daily Planet*, February 6. Available at <foodfirst.org/node/1621>.

Archer Daniels Midland Company. 2007. "ADM Statement Regarding Expanded Renewable Fuel Standard. Dec. 19." Available at <ethanolmarket.com/PressReleaseADM122007>.

Aslow, M. 2007. "Biofuels: Fact and Fiction." *Ecologist* 2, 19.

Baker, M.L., H.J. Dermot and B.A. Babcock. 2008. "Crop Based Biofuel Production Under Acreage Restraints and Uncertainty." Working Paper 09-WP 460. Center for

Agricultural and Rural Development, Iowa State University.

Birger, J. 2008. "The Ethanol Bust: The Ethanol Boom Is Running Out of Gas as Corn Prices Spike." *Fortune*, February 28. Available at <money.cnn.com/2008/02/27/magazines/fortune/ethanol.fortune/?postversi>.

Black, S. 2008. "Cargill Q1 Earnings Jump 62 percent." *Minneapolis/St. Paul Business Journal*, October 13. Available at <bizjournals.com/twincities/stories/2008/10/13/daily3.html?ana=from_rss>.

Bravo, E. 2006. *Biocombustibles, Cutlivos Energeticos y Soberania Alimentaria: Encendiendo el Debate sobre Biocommustibles.* Quito, Ecuador: Accion Ecologica.

Crutzen, P.J., A.R. Mosier, K.A. Smith, and W. Winiwarter. 2007. "Nitrous Oxide Release from Agro-Biofuel Production Negates Global Warming Reduction by Replacing Fossil Fuels." *Atmospheric Chemistry and Physics* 7.

Dufey, A. 2007. "International Trade in Biofuels: Good for Development? And Good for Environment?" IIED Briefing Papers. International Institute for Environment and Development, January. Available at <iied.org/pubs/display.php?o=11068IIED>.

Economist. 2007a. "Cheap No More." December 6.

_____. 2007b. "The World Goes to Town." May 11.

ETC Group. 2007. "Peak Oil + Peak Soil = Peak Spoils." ETC *Group Communique* N. 96 (November/December).

FAO (Food and Agriculture Organization of the United Nations). 2008a. "World Food Situation: High Food Prices." *Food Price Indices*, September 2008. Available at <fao.org/worldfoodsituation/FoodPricesIndex/en/>.

_____. 2008b. "Number of Hungry People Rises to 963 million." FAO *Newsroom*, December 9. Available at <fao.org/news/story/en/item/8836/>.

_____. 2008c. "World Food Situation." Last updated July 2008. Available at <fao.org/worldfoodsituation/wfs-faq/en/>.

_____. 2009. "Food Prices Remain High in Developing Countries." FAO Newsroom, April 23. Available at <fao.org/news/story/en/item/12660/icode/>.

FBOMS (Brazilian Forum of NGOs and Social Movements for the Environment and Development). 2006. "Agri-businesses and Biofuels: An Explosive Mixture." Rio de Janeiro.

FOEI (Friends of the Earth International). 2006. "Challenging Cellulose Industry: The Impacts of Pulping in South America." Briefing Paper for the People's Tribunal on Human Rights Violations, May 9. Available at <foei.org/en/publications/forests/Briefing_pulp_and_paper_projects.rtf>.

Federal Trade Commission (FTC). "Report on Ethanol Market Concentration." 2005. December 1. Available at <ftc.gov/reports/ethanol05/20051202ethanolmarket.pdf>.

Geertz, C. 1963. *Agricultural Involution.* Berkeley, CA: University of California Press.

GRAIN. 2007a. "The Sugar-cane–Ethanol Nexus." Available at <grain.org/seedling/?id=488>.

_____. 2007b. "Corporate Power: Agrofuels and the Expansion of Agri-business." *Seedling* July. Available at <grain.org/seedling_files/seed-07-07-3-en.pdf>.

_____. 2008. "Seized: The 2008 Land Grab for Food and Financial Security." GRAIN *Briefing*, October. Available at <grain.org/go/landgrab>.

Harvey, D. 2003. *The New Imperialism.* New York: Oxford University Press.

Hassan, H. 2007. "Overview of the US Ethanol Market." *Institute for Food and Development Policy.* July 24. Available at <foodfirst.org/en/node/1723>.

Holt-Giménez, E. 2007. *Land – Gold – Reform. The Territorial Restructuring of Guatemala's*

Highlands. Food First Development Report No. 16. Oakland, CA: Institute for Food and Development Policy.

_____. 2008. *The World Food Crisis: What's Behind It and What We Can Do about It*. Food First Policy Brief 16. Institute for Food and Development Policy. October 3. Available at <foodfirst.org/en/node/2264>.

Holt-Giménez, E., and R. Patel. 2009. *Food Rebellions! Crisis and the Hunger for Justice*. Oakland, CA: Food First Books.

Howarth, R.W., S. Bringezu, M. Bekunda, C. de Fraiture, L. Maene, L.A. Martinelli and O.E. Sala. 2009. "Rapid Assessment on Biofuels and the Environment: Overview and Key Findings. Executive Summary." In R.W. Howarth and S. Bringezu (eds.), *Biofuels: Environmental Consequences and Interactions with Changing Land Use*. Proceedings of the Scientific Committee on Problems of the Environment (SCOPE) International Biofuels Project Rapid Assessment, September 22–25, 2008, Gummersbach, Germany.

Hurtado, L. 2009. "Agrofuel Plantations and the Loss of Land for Food Production in Guatemala." In J. Richard (ed.), *Agrofuels in the Americas*. Oakland, CA: Food First Books.

Inhofe, J. 2007. "Inhofe Introduces President Bush's 'Alternative Fuel Standard Act of 2007.'" Press Release. U.S. Senate Committee on Environment and Public Works. April 19. Available at <tiny.cc/QbGBs>.

Jacobsen, M. 2009. "Review of Solutions to Global Warming, Air Pollution, and Energy Security." *Energy and Environmental Science* 2.

Jonasse, R. 2009. "Agri-business' Field of Dreams: IFIs and the Latin American Agrofuels Expansion." In R. Jonasse (ed.), *Agrofuels in the Americas*. Oakland, CA: Food First Books. Available at <foodfirst.org/en/node/2426>.

Lilley, S. 2004. "Paving the Amazon with Soy: World Bank Bows to Audit of Maggi Loan." *CorpWatch*. Available at <corpwatch.org/article.php?id=11756>.

Meirelles, D. 2005. "Papel para el Norte, Hiper Consumo de Agua en el Sur. Una Hidrogenealogía de las Fábricas de Celulosa de Aracruz." In Ortiz et al. (eds.), *Entre el Desierto Verde y el País Productivo*. Montevideo: REDES-AT – Casa Bertolt Brecht.

Monbiot, G. 2007. "If We Want to Save the Planet, We Need a Five-Year Freeze on Biofuels." *Guardian*, March 3.

Moreno, C. 2006. "Agroenergia X Soberania Alimentar: a Questão Agrária do século XXI." Plano Nacional de Agroenergia 2006–2011.

Morris, David. 2005. "Do Bigger Plants Mean Fewer Farmer Benefits?" *Rural Cooperatives* 72, 6 (November/December).

Morton, D.C., R.S. DeFries, Y.E. Shimabukuro, L.O. Anderson, E. Arai, F. del Bon Espirito-Santo, R. Freitas and J. Morisette. 2006. "Cropland Expansion Changes Deforestation Dynamics in the Southern Brazilian Amazon." *PNAS* 3.

O'Driscoll, P. 2005. "Part of the Problem: Trade, Transnational Corporations and Hunger." *Center Focus* 166. Washington, DC: Center of Concern.

Philpott, T. 2007. "Corn Ethanol to the Max." *The Gristmill*, December 12. Available at <gristmill.grist.org/story/2007/12/11/85259/793>.

Reeves, S. 2007. "Green Technology Revs Up Venture Capitalists." *CNBC Stock Market News*, March 6. Available at <cnbc.com/id/17130665>.

RFA (Renewable Fuels Association). 2010. "Biorefinery Locations." Available at <ethanolrfa. org/industry/locations/>.

Runge, F.C., and B. Senauer. 2007. "How Biofuels Could Starve the Poor." *Foreign Affairs*

May/June.

Scully, V. 2007. "Effects of the Biofuel Boom: Market Views." *Business Week,* August 27.

Searchinger, T., R. Heimlich, R.A. Houghton, F. Dong, A. Elobeid, J. Fabiosa, S. Tokgoz, D. Hayes, T.H. Yu. 2008. "Use of U.S. Croplands for Biofuels Increases Greenhouse Gases Through Emissions from Land-Use Change." *Science* 319: 5867.

Sen, A. 1981. *Poverty and Famines: An Essay on Entitlement and Deprivation.* Oxford, Clarendon Press.

Senauer, B. 2008. "The Appetite for Biofuel Starves the Poor." *Guardian,* July 3.

Shattuck, A. 2008. *The Agrofuel's Trojan Horse: Biotechnology and the Corporate Domination of Agriculture.* Food First Policy Brief 14. Oakland, CA: Institute for Food and Development Policy. Available at <foodfirst.org/en/node/2111>.

Tillman, D., and J. Hill. 2007. "Corn Ethanol Can't Solve our Climate and Energy Problems." *Washington Post,* March 25.

Ugarte de la Torre, D.G., and S. Murphy. 2008. "The Global Food Crisis: Creating an Opportunity for Fairer and More Sustainable Food and Agriculture Systems Worldwide." *EcoFair Trade Dialogue Papers* 11, October.

USDA (United States Department of Agriculture). 2008. "Adoption of Genetically Engineered Crops in the U.S." *USDA Economic Research Service.* Available at <ers.usda. gov/Data/BiotechCrops/alltables.xls>.

Vorley, B. 2003. "Food, Inc. Corporate Concentration from Farmer to Consumer." *UK Food Group.* Available at <ukfg.org.uk/docs/UKFG-Foodinc-Nov03.pdf>.

Weitzman, H. 2009. "Cargill Feels Chill of Downturn." *Financial Times,* April 15.

Zimbalist, Z. 2007. "Columbia Palm Oil Biodiesel Plantations: A 'Lose-lose' Development Strategy?" Food First Backgrounder. *Institute for Food and Development Policy* 13, 4.

7

Reconnecting Agriculture & the Environment

Food Sovereignty & the Agrarian Basis of Ecological Citizenship

Hannah Wittman

> Seed eaters have responsibilities. — Colin Duncan, *The Centrality of Agriculture*, 1996

In the face of challenges presented by climate change, food shortages and widespread environmental degradation, there is an increasing urgency to revisit the character of the changing relationship between society and nature. Arguing that today's precarious conditions are a direct result of the longstanding ideological separation between society and nature in models of production and social organization, advocates of "resilience" models of society/nature relations demand a change in how we think about what nature is and how it is to be used (Berkes et al. 2003). These ideologies of nature reflect not only the social and historical characteristics of individuals and societies, but also the ecological characteristics of the areas in which these societies are embedded (Redclift 2006).

In this chapter, I examine the ways in which agrarian communities with long-standing relationships and rights to the land have been disconnected from the ecological basis of citizenship by rural modernization strategies based on the separation of society from nature. The resulting dominant model of efficiency-based agriculture is now being challenged by a food sovereignty model founded on grassroots practices of agrarian citizenship and ecologically sustainable local food production, which reconnect agriculture, society and the environment through systems of mutual obligation. By working towards food sovereignty through concrete campaigns that involve practices of agrarian citizenship, rural communities promote a rationality of resilience and concern for the future of nature.[1]

Technology, Efficiency & the Separation of Nature from Agriculture

The history of agricultural modernization is a long one, but one of its fundamental premises is the notion of efficiency. Dating back to Adam Smith's theory of labour specialization, the history of efficiency is a curious one; as populations and economies struggle to grow in a landscape of finite natural materials and human labour, societies seek ways to "do more with less" through social innovation and the transformation of traditional modes of production. Thus, efficiency emerged following the Industrial Revolution as a particularly dominant principle of social organization (Princen 2005).

The contemporary logic of efficiency, which measures ratios of inputs (material resources, labour, knowledge) to output (goods and services), derives from a historical notion that encompassed multiple goals, ratios and logics to come together to produce particular outcomes. During the Industrial Revolution, the practice of separating, simplifying and specializing in distinct tasks and practices produced a more clearly defined division of labour. The concept of efficiency during this period was further applied as a social construction to promote industrialization, economic expansion and consumerism (Princen 2005: 53–55). The idea was that by reducing productive activities to their constituent parts, each one could be simplified and streamlined and thus more "output" could be gained from each unit of input — materials or labour.

Efficiency as a basis for a modern economic rationality thus became a model for agricultural production. With the invention of the internal combustion engine and innovation in affordable gas-powered farm implements at the turn of the twentieth century, the ability to produce more food, faster and with less labour, became a reality. Other mechanical innovations followed, allowing a formidable landscape transformation as cultivation expanded to meet growing demands for food and inputs for other industrial activities. For example, at the end of the nineteenth century, the production of a hundred bushels of corn required thirty-five to forty hours of planting and harvesting labour. Today, with the use of large tractors for ploughing, weeding and harvesting and the application of chemical fertilizers, this field time has been reduced to less than three hours (Constable and Somerville 2003). With the rapid increase in corn production, industries found new ways to use the now more cheaply available product for further industrial innovation (Pollan 2006).

On the ecological side, efficiency-driven simplification and standardization involved such practices as reducing the number of seed varieties used for major cereals and reducing the diversity of agricultural landscapes (e.g., mono-cropping). The simplification of the landscape made mechanized planting and harvesting easier and allowed for the widespread application of chemical fertilizers and pesticides. By the mid twentieth century, the Green Revolution exported these agricultural innovations to the developing world, seeking to improve world food availability by increasing productivity and streamlining an industrial model of production. Indeed, during the 1990s alone, the production of cereal crops increased by 17 percent, roots and tubers by 13 percent and meat by 46 percent (World Resources Institute 2000, cited in McNeely and Sherr 2002).

However, these social and technological innovations also had many negative ecological effects. The reduced genetic variability required by large-scale, mechanized agriculture can decrease resistance to pests/predators. Extensive mono-cropping also requires high levels of chemical inputs, which can lead to soil degradation, desertification and water pollution. The destabilization of ecosystems linked to this industrial model of agriculture not only reduces the stability of food production but also reduces the resilience of ecosystems in the face of climate change and other disturbances (Kareiva et al. 2007).

Conventional, export-oriented economic development strategies and Green Revolution technologies have also had social consequences, including increased poverty levels and a failure to benefit farmers in developing countries in particular. Farmers worldwide continue to lose access to land and productive capacity as a result of agricultural restructuring and the increasing mobility of industrialized agriculture, making them increasingly dependent on imported or purchased inputs and food. Despite massive increases in food production that have, to date, outpaced human population growth, today over a billion people go hungry due to increasing poverty and the inability to purchase food (FAO 2009). In addition, studies show that food losses and waste in the journey from "field to fork" in the contemporary industrial food system, which relies on regional specialization and long-distance transport of food, may exceed 50 percent (Lundqvist et al. 2008). Distribution issues contribute not only to the persistent problem of hunger but also to climate change via the greenhouse gas contributions of transportation networks and emissions related to excess production.

Smallholder agriculture in the developing world has struggled to compete with subsidized imports from North American and European countries, which operate on a model that equates efficiency with yield and depends on a long history of state support for export infrastructure and production credits. These "inputs," including subsidies on petroleum and on the development of particular agricultural technologies that also have high social costs (e.g., genetic modification), are not calculated in the input/output ratios of contemporary efficiency models. Other externalities, including the social welfare costs incurred by rural displacement and the loss of ecological services caused by mono-cropping, are also not calculated against the high yields of a single-crop model.

National agriculture policies aim to improve "aggregate" production, awarding subsidies to those crops deemed most important and efficient for international trade. The modern industrial agricultural system also requires a regulatory framework in which farmers are identified as individual and competitive business managers rather than as a collectivity of earth stewards. The increased bureaucratization of the industrial farming model has further excluded those farmers who have been unable to access land titles, bank accounts or futures markets. Even procedures as common as recording units of milk production have served to sever the human-animal relationship, creating practices based on "individual units of production" and separating out the constituent parts of a farm enterprise rather than seeing it as a whole (Nimmo 2008).

As less labour was required for large-scale food production systems, the rural-urban interface changed. Following a cycle experienced in Europe in the nineteenth century, in which enclosure pushed former agricultural labourers to the cities, many rural communities in late twentieth-century North America became ghost towns surrounded by vast acreages of highly mechanized, hybrid technology corn and soybeans. Much as predicted by modernization theorists claiming a "disappearance of the peasantry," in the U.S. and Canada the percentage of farming families has dropped from over 90 percent of the population at the beginning of the nineteenth century to under 1 percent today. In Brazil, over five million peasants have been driven off the land since the late 1970s. A global rural exodus fostered by the industrialization of agriculture and the drive for efficient large-scale production has increased the distancing of humans from their food source (Pretty 1995). The increasing loss of indigenous knowledge and diversity in production methods has also reduced the capacity of traditional agricultural stewards to manage productive landscapes.

To sum up, the modernization of agriculture is founded on a simple, economic ideology of nature, which in turn depends on the utilization of certain technologies in the name of efficiency and the expansion of capital-oriented production. This transformation of agrarian social and ecological conditions has served to disrupt agriculture as a holistic link between human culture and the environment, producing a chasm, or "metabolic rift," between humans and nature (Foster 1999, 2000; Moore 2000; Wittman 2009b). Agriculture's historically relatively closed-loop system (food production and reincorporation of wastes into the traditional agrarian cycle) is disrupted as producers and consumers are increasingly separated, not just in the division of rural/urban spaces but also further afield through agricultural trade and regional specialization. This process of distancing underlies and fosters the systematic effects of agricultural restructuring and its particular implications for both society and nature.

Agrarian Citizenship as an Alternative Agroecological Rationality

The imposition of a modern economic rationality has not fully erased, however, the existence of agricultural frameworks that prioritize the maximization of ecological and social capital rather than economic output. As Princen stresses, alternatives based on the principles of ecological rationality and resilience assume that the

> common good, especially with respect to the "commons" of global climate stability and biodiversity, of local clean water and fertile soil, cannot be achieved by simply aggregating individual private "goods".… An ecologically rational society conditions individual decisions so as to be attentive to ecosystem fragility and to enhance ecosystem resilience. (2005: 25–26)

Considerations of ecological citizenship pay attention to the ways that individuals and communities assess their ecological impact in a cosmopolitan world of scarce resources, focusing in particular on the political communities that constitute new regulations and practices that protect the environment (Dobson 2003; Dobson and Bell 2006). The concept of agrarian citizenship takes this one step further, recognizing how the political and material rights and practices of rural dwellers are integrated into the socio-ecological metabolism between society and nature (Wittman 2009a, 2009b). Agrarian citizenship thus recognizes the roles of both nature and society in the continuing political, economic and cultural evolution of agrarian society.

Radical resistance to the expansion of market capitalism, based on a neoliberal ecological rationality, involves reshaping social and ecological relations. This reconstitution of both society and nature responds and adapts to social and ecological crisis. It is based on the movement of active social and ecological processes reacting to the excesses of an increasingly globalized market, which depends on the ongoing separation between individuals, society and the ecological basis of reproduction. This new linkage of urban and rural ecological and agrarian citizens, acting in response to and in concert with nature, is founded upon a reconfigured notion of the rights and responsibilities between humanity and nature and a revised ecological rationality. Actively reconstituting themselves in the face of market pressure, modern agrarian citizens thus reclaim the notion of a humanistic community that not only demands state re-regulation of the market but also acts to protect itself against the continued decimation of social and ecological spaces.

The contemporary landscape of worldwide peasantries and the changing modes of agricultural production offer several examples of this alternative framework, which challenges the dominant neoliberal trade model and offers evidence of re-emerging practices of agrarian citizenship. Despite the displacing effects of the industrial transformation of agriculture, just under half of the world's population still lives and works in rural areas, with agricultural households still constituting about two-fifths of the world population (Weis 2007). In addition, agricultural movements worldwide including La Vía Campesina are challenging the global role of industrial capital and agri-business.

In what follows, I provide an overview of some of the specific ways in which members of contemporary peasant movements in Brazil, as members of La Vía Campesina, work towards food sovereignty by placing a particular emphasis on the role of ecology in the relationship between environment, society and citizenship. Responding to specialization with diversification, to efficiency with sufficiency and to commoditization with sovereignty, these contemporary alternative agricultural movements are challenging the corporate and neoliberal trends of agrarian transformation towards a more sustainable future.

Food Sovereignty: Enacting Agrarian Citizenship

We cultivate the earth and the earth cultivates us. — MST Mistica 2003

In La Vía Campesina's terms, a people's food sovereignty encompasses the right of local populations to define their own agricultural and food policy, to organize food production and consumption to meet local needs and to secure access to land, water and seed (La Vía Campesina 2007). As an organizing principle, food sovereignty covers a variety of related aims and action areas that involve the unification of social, environmental and agricultural principles. Explicit attention to the human/ecology link is present in most of La Vía Campesina's eight priority issue areas (agrarian reform, biodiversity and genetic resources, food sovereignty and trade, sustainable peasants agriculture, migration and rural workers, gender, human rights and youth). The first five of these areas directly address the challenges posed to the environment by the industrialization of agriculture in the name of economic globalization.

La Vía Campesina (2006) stated that "it will continue the struggle in defence of natural resources, for land, water, seeds, forests, life and for the protection, care and improvement of biological and cultural diversity." To do so, the movement fosters the universalization of a particular discourse and set of principles around globalization and the environment and the proposal of a variety of concrete alternatives. These alternatives are not aimed at streamlining a particular economic rationality (in effect replacing one standard model with another) but advocate the principle of "unity within diversity." This approach resembles what Princen (2005) calls "rational pluralism," or the principle of protecting biological and cultural diversity (in effect, protecting nature's right to act) while maintaining a direct social connection to the land and environment, in this case through agricultural production. For example, one Brazilian activist argues:

> The process of modernization increased the pace of production and reduced the variety of foods. This didn't offer quality to the human being, but on the contrary brought grave problems. It is this sense that La Vía Campesina wants to bring back and wants to construct an agriculture that doesn't just preserve the environment, but it also produces quality and healthy food. (member of Brazil's Pastoral Land Commission, July 2006)

This activist's critique of modern agriculture is founded on the practice of agrarian citizenship, the idea that rural producers have not only rights to the land and the environment but also responsibilities, connected to these rights, for maintaining the diversity of socio-ecological reproduction. In his critique of an industrial agricultural model that views itself as "the only model possible," this activist continues:

This market doesn't place responsibility on the producer. So you produce without any responsibility for what it will cause for the environment, for people, for the general population. It's a model that we could say is totally irresponsible.

By rationalizing production systems and evaluating outcomes solely on the basis of tradability and yield, as in the industrial agricultural production model, food becomes solely a market commodity and loses its characteristics that would have other values in a system based on ecological rationality. These include cultural diversity, agro-biodiversity and nutritional values and are important parts of what members of La Vía Campesina see as a "harmonious" relationship with nature, which involves a relation of mutual obligation rather than domination. The activist continues:

> Men, women, human beings [now live] as if they had dominion over nature. We need to consider that other elements also exist within nature, and for that reason nature must be respected. We need to have a harmonious relation because nature needs humans and humans need nature.

Within Brazil, the campaign against green deserts and the seed sovereignty campaign, both initiatives of La Vía Campesina's Biodiversity and Genetic Diversity Working Group, are particularly illustrative of how contemporary peasant movements are engaging in a practice of agrarian citizenship aimed at reconnecting and repairing the relationship (and in a sense re-establishing equilibrium) between human society and nature. These campaigns highlight the importance of biodiversity in agricultural systems and of supplying local needs before filling international trade markets. They also express what food sovereignty looks like in a model of agriculture in which both nature and society are protected in their rights to independently exist and evolve.

The Campaign against Green Deserts

On International Women's Day, March 8, 2006, over two thousand La Vía Campesina members from several countries entered a eucalyptus plantation in the Southern state of Rio Grande do Sul, Brazil, owned by the multinational corporation Aracruz. Aracruz plantations cover over 250,000 hectares in Brazil, with over fifty thousand hectares in Rio Grande do Sul. After symbolically destroying a set of genetically modified tree seedlings in a plantation greenhouse, the women issued a manifesto stating:

> We are against green deserts, the enormous plantations of eucalyptus, acacia and pines for cellulose, that cover thousands of hectares in Brazil and Latin America. When the green desert advances, biodiversity is destroyed, soils deteriorate, rivers dry up. Moreover cellulose plants pollute air and water and threaten human health. (Manifesto Women of La Vía Campesina, March 8, 2006)

The members of La Vía Campesina were careful to indicate that their issue was not with the eucalyptus plant itself; even though eucalyptus originated in Australia, it had been grown in Brazil for decades. As pointed out in an informational pamphlet about green deserts in Brazil:

> For many years, eucalyptus didn't create problems for the environment in Brazil, as farmers planted one here, another there, in small woodlots to have firewood that grew more rapidly or even for lumber. As they were planted in small quantities and well distributed in a large geographic space, they brought the benefit of producing good fuel wood in little time and almost no problems for mother nature, for the environment. So it was that the eucalyptus spread out through various regions of Brazil, for many years without creating problems for anyone and even bringing various benefits. (La Vía Campesina-RS 2006: 7)

The term "green desert" emerged in the 1970s in the Brazilian states of Espiritu Santo and Minas Gerais, where a local priest was concerned with the rapid expansion of eucalyptus plantations used by cellulose paper industries and for the production of charcoal used by Brazil's pig-iron industry, both priorities in the industrialization program of Brazil's military government. The Pastoral Land Commission of Brazil, long associated with rural farmers' unions, became concerned that not only were these plantations crowding out traditional smallholders and local food production, but the water-hungry eucalyptus plantations were also having detrimental effects on the water table and surrounding ecology. In short, it was the way in which human institutions (the market, industrial policy and capitalism) engaged with eucalyptus production that was causing the ecological problems. The rationalization of production through large-scale mono-cropping, in the name of comparative advantage and efficiency, led to what rural activists called a green desert.

One Brazilian Pastoral Land Commission activist explained that the green desert concept captured the particular ways that large-scale eucalyptus plantings and other non-native species such as American pine were lowering water tables and reducing wild biodiversity:

> You have a tree that is pretty, green, but doesn't have life. This is why the concept of green desert was created because in truth it's green, but it eliminates life, not entirely, but it eliminates fundamental parts of life, that are the animals and water.

The activist added that much of today's eucalyptus production in Brazil is carried out on a contract basis, where large processors encourage small farmers to expand their eucalyptus production:

> The multinational companies don't want to run the risk of producing the eucalyptus. They don't want to run the risk of the grave environmental problems

that will come out of the green desert.... They exempt themselves from the responsibility to produce this product, to cause damage to the environment and to the owners of that space. Does the world need that much paper, that much cellulose?

The concept of the green desert has been rapidly globalized by members of La Vía Campesina in locally based campaigns against agricultural monoculture and genetically modified organisms, thus effectively constructing a universal discourse around the need to protect water, genetic resources and land as the essential elements of all life. Agronomist Carlos Alberto Dayrell, of the Centre for Alternative Agriculture in Minas Gerais, Brazil, remarked:

> This act of La Vía Campesina women should be understood as an alarm bell regarding the risks of a developmentalist option that compromises the possibility of a future for all of us Brazilians, for all of us citizens and human beings of the world. (Cassol 2006)

The campaign against green deserts, as a direct call for actions founded upon agrarian citizenship, was also advanced at the conferences of the Parties to the Convention on Biological Diversity in 2006 and 2008. At these meetings, food sovereignty advocates highlighted the connection between seed diversity and green deserts and their relationship to the future of global food systems.

The Seed Sovereignty Campaign

> "The farmers love biodiversity. It guarantees them seeds and life." —La Vía Campesina Mistica, COP-9, Bonn[2]

La Vía Campesina and member organizations in Brazil continue to be particularly vocal and active on international food policy agendas around the topics of seed sovereignty and the protection of global agro-biodiversity. Their critique is centred upon the monopolization of genetic resources by multinational corporations, which has, in their view, threatened the loss of heritage seed varieties adapted to local landscapes and ecological conditions. In an era of climatic uncertainty, access to biologically resilient seeds that can be utilized under changing conditions is critical for ensuring not only local food security but also for maintaining a foundation for agroecological resilience. By recognizing the importance of biological and genetic diversity, the seed sovereignty campaign seeks to ensure the viability of local seed varieties while also embracing peasant innovation in diversifying agricultural landscapes. Regarding their seed initiatives, one Brazilian MST activist commented:

> The principle is to cooperate in a flexible way, so that everyone has access to forms of cooperation. We are searching for technological change in the area of

seeds. This isn't just looking for seeds from our grandparents' past just because it's a seed that our grandparents planted. We are looking to improve that seed, not within the concept of change related to agro-chemicals or transgenics, but within a technological change that is productive, that is honourable for the producer.... What's important is that these seeds and this agricultural productivity aren't in the hands of a company that guards these seeds, that guards this historical archive, but that they're taken by the population, that they're produced by the population in millions and that a million peasants can reproduce this technology so that it's at the service of agrarian reform and not for some companies to earn money later. It's to be *absorbed* by the people. We already have several varieties of seeds, many of them have a much higher productivity than the agro-chemical ones. And they cost less.

As part of the campaign, La Vía Campesina has participated in the last two meetings of the Convention on Biological Diversity (COP-8 in Curitiba, Brazil, in 2006 and COP-9 in Bonn, Germany, in 2008). In both instances, La Vía Campesina and its allies contested the concentration of seed ownership, primarily in the hands of multinational agri-business, for its implications not only for food sovereignty and the ability of local peoples to have access to appropriate seed stock but also for the ecological implications of genetic modification of seeds. By directly challenging "Terminator" technology, which disrupts the biological ability of a seed to reproduce itself and thus be saved for use by subsequent generations — the very foundation of a socio-ecological metabolism — groups of peasants presented a clear example of the obligations of agrarian citizenship: to protect the future resiliency and productive capacity of agriculture.

From March 13–27, 2006, representatives of La Vía Campesina gathered in the COP-8 meeting in Curitiba, Brazil, to carry out a series of protests, marches and educational events culminating in the Global Forum of Civil Society to debate issues around globalization, biodiversity and civil-society response to the COP negotiations. João Pedro Stédile, a national spokesperson for Brazilian Landless Workers' Movement (MST), made a link between the Aracruz case and the COP-8 negotiations: "Aracruz is a green desert. There is no life there, and not even the bees survive in the plantations of the supposedly advanced Aracruz" (cited in *O Estado de São Paulo* 2006). At the global civil-society forum, MST and La Vía Campesina members extended this green desert analysis to the cultivation of all genetically modified seeds, especially soybean, which they see as a direct threat to both the environment and rural livelihoods.

A delegation of representatives from La Vía Campesina also met with the executive secretary of the Convention on Biological Diversity (CBD) to relate their concerns about the possible lifting of the CBD moratorium on Terminator seeds, which had been in place since 2000, just two years after the first patents on these seeds were issued. Francisca Rodriguez, a La Vía Campesina representative from Chile, said at that meeting: "We will not stop until Terminator disappears from the face of the earth," an appeal that was ultimately successful as the moratorium was upheld despite opposition from the U.S., Canada, New Zealand and Australia. A La Vía Campesina press release of March 27 stated: "The rejection of the 'suicide seeds' is an essential step for securely implementing the proposals on agro-biodiversity, biodiversity, and food sovereignty discussed in the COP-8" (cited in Sharratt 2006).

Considering seeds as "the patrimony of humanity" brings us back to the question of how agriculture and agrarian citizens are involved in a metabolic relationship with nature, one involving the preservation of rights and responsibilities for each party (Wittman 2009b). Seeds are biological phenomena — plants, by their nature, produce seeds. These seeds are made available for food (e.g., cereal grains) and are also stored and planted for further use as seed stock and food. Demands by peasant activists for seed biodiversity to remain freely available to humanity recognizes the sovereignty of both people and nature in having access to this seed collective, rather than it being enclosed and commoditized under the control of a limited number of corporations. The idea of reciprocity between humanity and nature is especially evident here, as the seed biodiversity of food crops has expanded tremendously through human actions over the past millennia, even as the number of commercially grown species continues to diminish under the industrial food regime. One Brazilian activist comments:

> It's in the name of "humanity" that imperialist governments, multinationals appropriate biodiversity and anything that can be turned into merchandise for themselves…. We're talking about humanity in a different sense. We're talking about all the human beings on planet Earth, of future generations. We're also taking about taking care of those things that *can't* be appropriated by one person, by one corporation.

The dissemination and preservation of a large variety of locally adapted seeds for the collective good of both people and nature thus disrupts the notions of specialization and commoditization that are so fundamental to the dominant model of agricultural production. In recognizing the rights of nature to exist and be protected from both commoditization and degradation through overexploitation, La Vía Campesina and its counterpart, the MST, are practising an agrarian citizenship that calls for the re-engagement of small-scale producers with social and ecological responsibility, not only to their local communities but to the local ecology. Orienting production towards the needs of local communities via a social contract of responsibilities to the public good — whether that is establishing eucalyptus plantations for local paper production, developing local seed networks or returning export-oriented tracts to productive orientations for local use — is a foundational aspect of food sovereignty.

Conclusion

Agrarian citizenship, a concept that links agricultural practice to environmental and social sustainability, is a key component of the theoretical framework of food sovereignty. Importantly, agrarian movements associated with La Vía Campesina distinguish between the citizenship obligations of agroecological production, and other "moralizations" of production within the dominant capitalist model. For example, the dominant model of environmental management through bio-diversity reserves, national parks and conservation agriculture all result in the removal of land from food production, still separating society from nature but now placing that separation on a moral basis. This can be accomplished, so this formulation goes, through conservation initiatives based on resource-preservation via separation. In this moralized preservationist framework, the "protection" of nature from humans creates a rising number of "conservation refugees," a term referring to local inhabitants and farmers who are excluded from traditional, and often exclusionary, national park models, which in the long-term often fail to be "preserved" (Geisler and De Sousa 2001; Geisler 2003). In addition, millions of small farmers have been displaced by the promotion of so-called "eco-agriculture" or land-sparing initiatives. These as-yet-unproven models promote the increased use of genetically modified seeds and other agricultural modernization techniques with the objective of obtaining higher agricultural yields in order to "save land for nature" in environmental reserves and national parks (McNeely and Sherr 2003; Green et al. 2005).

The alternative is a stewardship model in which land is sustainably managed for multiple objectives that include the production of food and the protection of biodiversity. In this regard, members of La Vía Campesina in Brazil don't view themselves as part of the traditional environmental movement; nor are they commonly included in analysis of transnational environmental organizations, many of which continue to uphold the separation of nature and society emblematic of the conservationist model described above. However, bringing environmental considerations — as citizenship obligations — into the agrarian struggle is a key part of the transition to a more synergistic model of socio-ecological relations. A long time MST activist explains:

> It's not just defending the little butterflies; we have to defend life in its entire sense of being alive. So I think that this elevates the debate and also ends up provoking those organizations that are more specifically environmental. It doesn't work for me to just defend a species, a plant, a tree or a wild animal. If you don't think about the whole and don't fight against the multinationals, against the economic model, against imperialism and its sense of domination over the peoples, economic and cultural domination, you don't have a clear position against the war.

For La Vía Campesina and the MST, food sovereignty thus reflects a broad vision of an agrarian citizenship that values both society and nature, and as such it is a more holistic model since it does not protect one at the expense of the other. A socio-ecological peasant movement based on a practice of agrarian citizenship that is at once resilient, regenerative and responsible effectively provides an alternative to the ongoing debates over which model of "land sparing" and "wildlife-friendly" agriculture is most likely to preserve biodiversity (Berkes 2004; Vandermeer and Perfecto 2007; Fischer et al. 2008).

The concept of agrarian citizenship thus encompasses the rights of nature and humans to collectively produce food for community sustenance, alongside a mutual responsibility to uphold the future productivity of the land. In order to protect these rights, La Vía Campesina and the MST see a role for environmental regulation in the construction of new relations between society, community, land and nature. Members argue that they confront the issues of environment and globalization on two fronts: one, ensuring that local and national state actors regulate corporations to respect environmental laws, which may include participating in international treaties and protocols such as COP-8 and Kyoto; and two, enacting concrete alternatives to neoliberal globalization at the local level.

Through campaigns such as the occupation of the Aracruz seedling plantation and the formal meetings at the Convention on Biological Diversity, members of the MST and La Vía Campesina bridge the local and the global by globalizing the discourse of the green desert to protest the multinational causes of some of Brazil's environmental problems and providing concrete examples of a socio-ecological alternative. Rather than pushing a particular agenda of simply regulating the environmental behaviour of individuals and corporations through international treaties, La Vía Campesina and the MST pursue a broader agenda of active social transformation. These forces use the methods of protective legislation and instruments of intervention such as land reform, land laws and agrarian tariffs to protect natural resources and culture of the countryside. They also aim for constitutional reform, fair trade, community-based intellectual property rights for indigenous knowledge (a pillar of seed sovereignty), agroecology and farmer-based research. In all of these elements, a change in environmental behaviour is part and parcel of a new society based on the universal principles of social justice, land and food sovereignty and respect for life. With a principle of social organization based on ecological holism, the alternative production paradigm reconnects agriculture and environment through the application of new societal aims: the ultimate conservation of humans and nature by enfranchised agrarian citizens who value both the social and ecological functions of land.

Agrarian citizens, as demonstrated in the practices of groups like La Vía Campesina, enact horizontal relationships within and between communities (social capital) and local ecologies (ecological capital) as well as connecting vertically with broader communities encompassing "humanity" and the "environment." As such, they provide the building blocks of an alternative practice of agrarian citizenship that analytically bridges the local, global, social and ecological to contest the contemporary relations between economic globalization and environmental change. Rather than privileging any one of these lenses to see how it affects the other, this framework recognizes the mutual constitution of human and ecological responses to economic globalization.

Notes

This chapter is a revised version of "Reworking the Metabolic Rift: La Vía Campesina, Agrarian Citizenship, and Food Sovereignty," which appeared in *Journal of Peasant Studies* 36, 4 (2009).

1.   The argument is based on conclusions drawn from ethnographic and interview-based fieldwork conducted by the author in collaboration with Brazil's Movimento dos Trabalhadores Rurais Sem Terra (Rural Landless Workers Movement, MST) and with member organizations of La Vía Campesina-Brazil between 2003–06.

2.   Available at <http://www.youtube.com/watch?v=58alIJSyCS0>.

References

Berkes, F. 2004. "Rethinking Community-Based Conservation." *Conservation Biology* 18, 3.

Berkes, F., J. Colding and C. Folke. 2003. *Navigating Social-Ecological Systems: Building Resilience for Complexity and Change.* Cambridge: Cambridge University Press.

Cassol, Daniel. 2006. "A ousadia das mulheres prevalesceu." *Jornal Brasil de Fato.* May 2. Available at <www3.brasildefato.com.br/v01/agencia/especiais/desertoverde/news_item.2006-05-02.9815840275/?searchterm=camponesas>.

Constable, G., and B. Somerville. 2003. *A Century of Innovation: Twenty Engineering Achievements that Transformed Our Lives.* Washington, DC: Joseph Henry Press.

Dobson, A. 2003. *Citizenship and the Environment.* Oxford: Oxford University Press.

Dobson, A., and D. Bell (eds.). 2006. *Environmental Citizenship.* Cambridge, MA: MIT Press.

Duncan, C.A.M. 1996. *The Centrality of Agriculture: Between Humankind and the Rest of Nature.* Montreal: McGill-Queen's University Press.

FAO (Food and Agriculture Organization of the United Nations). 2009. *The State of Food Insecurity in the World 2009.* Rome.

Fischer, J., B. Brosi, G.C Daily, P.R Ehrlich, R. Goldman, J. Goldstein, D.B Lindenmayer, A.D Manning, H.A Mooney, L. Pejchar, J. Ranganathan and H. Tallis. 2008. "Should Agricultural Politics Encourage Land Sparing or Wildlife-Friendly Farming?" *Frontiers in Ecology and the Environment* 6, 7.

Foster, J.B. 1999. "Marx's Theory of Metabolic Rift: Classical Foundations for Environmental Sociology." *American Journal of Sociology* 105, 2.

_____. 2000. *Marx's Ecology: Materialism and Nature.* New York: Monthly Review Press.

Geisler, C. 2003. "A New Kind of Trouble: Evictions in Eden." *International Social Science Journal* 55, 1.

Geisler, C., and R. De Sousa. 2001. "From Refuge to Refugee: The African Case." *Public Administration and Development* 21, 2.

Green, R.E., S.J. Cornell, J.P.W. Scharlemann and A. Balmford. 2005. "Farming and the

Fate of Wild Nature." *Science* 307.

Kareiva, P., S. Watts et al. 2007. "Domesticated Nature: Shaping Landscapes and Ecosystems for Human Welfare." *Science,* 316, 1866–69.

La Vía Campesina. 2006. "Statement from La Vía Campesina in Support to the Women from Rio Grande do Sul, Brazil." Press release.

_____. 2007. "The International Peasant's Voice." July. Available at <viacampesina.org/main_en/index.php?option=com_content&task=blogcategory&id=27&Item id=44>.

La Vía Campesina-RS. 2006. "O Latifúndio dos Eucaliptos: Informações Básicas sobre as Monoculturas de Arvoroes e as Indústrias de Papel." Pontocom Gráfica e Editora Ltda, La Vía Campesina-Rio Grande do Sul.

Lundqvist, J., C. de Fraiture and D. Molden. 2008. "Saving Water: From Field to Fork: Curbing Losses and Wastage in the Food Chain." SIWI Policy Brief. Stockholm: SIWI.

McNeely, J.A., and S.J. Sherr. 2003. *Ecoagriculture: Strategies to Feed the World and Save Biodiversity.* Washington, DC: Island Press.

Moore, J.W., 2000. "Environmental Crises and the Metabolic Rift in World-Historical Perspective." *Organization and Environment* 13, 2.

Nimmo, R. 2008. "Auditing Nature, Enacting Culture: Rationalisation as Disciplinary Purification in Early Twentieth Century British Dairy Farming." *Journal of Historical Sociology* 21, 2/3.

O Estado de São Paulo. 2006. Nacional, March 15.

Pollan, M. 2006. *The Omnivore's Dilemma: A Natural History of Four Meals.* New York: Penguin Press.

Pretty, J.N. 1995. *Regenerating Agriculture: Policies and Practice for Sustainability and Self-Reliance.* London: Earthscan.

Princen, T. 2005. *The Logic of Sufficiency.* Cambridge, MA: MIT Press.

Redclift, M. 2006. *Frontiers: Histories of Civil Society and Nature.* Cambridge, MA: MIT Press.

Sharratt, L. 2006. "COP8: Terminator Moratorium Upheld!" *Etcetera Blog,* March 26. Available at <etcblog.org/2006/03/26/cop8-terminator-moratorium-upheld/>.

Vandermeer, J., and I. Perfecto. 2007. "The Agricultural Matrix and a Future Paradigm for Conservation." *Conservation Biology* 21, 1.

Weis, T. 2007. *The Global Food Economy: The Battle for the Future of Farming.* London: Zed Books.

Wittman, H. 2009a. "Reframing Agrarian Citizenship: Land, life and Power in Brazil." *Journal of Rural Studies* 25.

_____. 2009b. "Reworking the Metabolic Rift: La Vía Campesina, Agrarian Citizenship, and Food Sovereignty." *Journal of Peasant Studies* 36: 4.

8

Food Sovereignty & Redistributive Land Policies
Exploring Linkages, Identifying Challenges

Saturnino M. Borras Jr. & Jennifer C. Franco

The convergence of different crises — financial, food, energy and environmental — in 2008 sharpened the divide between two perspectives: a crisis *in* the system versus a crisis *of* the system. The fundamental position presented by La Vía Campesina, today's politically most important transnational agrarian movement (TAM), is the latter. For La Vía Campesina, the point is to change the world — expose and oppose the current dominant model of rural development and offer a viable global alternative. La Vía Campesina has developed and is in the process of fine-tuning and operationalizing its notion of "food sovereignty" as an alternative to the current global system.

Bringing the vision of food sovereignty down to the complex, messy real world exposes multiple dilemmas and challenges, for instance, with regard to the land dimension of food sovereignty. Consider the following hypothetical scenarios.

> A community of sugarcane workers in the state of São Paulo in Brazil is convinced that food sovereignty provides the only viable future for them. The workers want to produce some food to ensure their community food supply and some cane to raise some cash for their other needs. But they do not own the sugarcane plantation where they are hired seasonally to cut canes.

> A community of poor farmers in Mozambique is convinced that food sovereignty has the potential to offer better livelihoods and future to them. The farmers want to transform their community land into a vibrant food-oriented production hub to feed their community and the nearby working people in Maputo. However, the national government, local government and the local traditional chiefs have a different idea: they want to lease the same land to a European investor to plant *jatropha* to produce biofuels for cars in Maputo and Europe.

> A community of indigenous people in an upland community in Southern Philippines is inspired by the vision of food sovereignty. But their indigenous community land is officially classified as state-owned. The national govern-

ment is closing a long-term lease arrangement with the Chinese government to produce food and biofuels from this land for export to China. The indigenous people are being recruited to become workers in the venture.

In all three hypothetical situations, it is difficult to imagine how any initiative towards food sovereignty can take off when the community pushing for such an alternative vision has no effective control over land resources, and those who have the control (the elite — state and non-state) have visions of development fundamentally opposed to food sovereignty. Unfortunately, the hypothetical cases described here continue to be common in developing countries.

In most settings in the developing world today the first step towards food sovereignty must be a reform of land-based social relations to enable the rural poor to have access to and effective control over land resources. In the context of renewed interests in land by global financial, agri-business and chemical companies to produce food and agrofuels, the food sovereignty project becomes increasingly more difficult to operationalize. The global elite's widespread efforts to grab lands in the Global South coincide with the large-scale neoliberal efforts at reorienting land policies worldwide towards private property and the free market. These efforts are partly inspired by the thesis of Hernando de Soto, who argued that the world's poor remain in poverty because their most important asset, i.e., land, cannot be financially transacted in the market due to their lack of formal private land titles; hence, the advocacy for the formalization and privatization of land rights (De Soto 2000, but also see Nyamu-Musembi 2007). Privitization efforts are also promoted among international financial and development institutions, e.g., the World Bank's market-led agrarian reform (Borras, Kay and Lahiff 2008). Ultimately, food sovereignty is about effective control over wealth and power. The land dimension is perhaps one of the most difficult questions within the food sovereignty project.

In settings where land-based wealth and power are concentrated in the hands of domestic and global elites, without resolving the question of land property relations in favour of rural poor communities, food sovereignty will not be able to take off at all. Where the rural poor have pre-existing access to and relative control over land resources but the domestic and global elites are aggressively promoting models of development different from and even hostile to food sovereignty, such access to land resources does not automatically lead to food sovereignty-based development. Just as it is important to locate discussions about food sovereignty within the broader context of the agrifood complex and food regimes, it is fundamental to our understanding of food sovereignty to locate it within the broad debates about land policies.

This chapter provides a mapping that can help specify which types of land policies are hostile to food sovereignty and which are not. We build on and add to the initial discussions about this topic (see, for example, La Vía Campesina 2008;

Rosset 2006; Rosset, Patel and Courville 2006; and Monsalve et al. 2006) and identify difficult challenges and terrains of political struggles over land.

Land-Based Social Relations, Not "Things"

It is important to clarify a few interrelated concepts and issues about property rights and land policies. First, by "ownership and/or control over land resources," we mean effective control over the nature, pace, extent and direction of surplus production, distribution and disposition. This framing enables us to detect actually existing land-based social relations regardless of what official documents claim. For example, much public land in Brazil, which official government records show is "empty" and not privately owned, is under production in informal, and often illegal, landlord-tenant relationships. In Indonesia, 70 percent of all lands are officially categorized as public forest land, most of which is assumed to be without inhabitants and productive activities, despite complex production relations within existing communities or between peasants and private companies engaged in oil palm plantations (Bachriadi 2009: 3). This framing also provides us with a disaggregated view of the competing social classes linked to each other by their varying relationships to land. Second, a land policy does not emerge from, nor is it carried out in, a vacuum. When carried out in the real world, a land policy causes a change in the actually existing land-based social relations. Some changes favour the landed classes, others elites or the state, while others may favour the poor. Third, land laws and land policies are not self-interpreting nor self-implementing (Franco 2008). It is during the interaction between various, often conflicting, actors within the state and in society that land policies are actually interpreted, activated and implemented in a variety of ways from one place to another over time. Fourth, land-based social relations are varied and diverse from one setting to the next and are shaped by socio-economic, political, cultural and historical factors. Fifth, land-based social relations are dynamic and not static. These are not like development projects, which can be contained within a timeline. Land-based social relations may continue to be dynamically altered long after a land titling project or a land reform program has officially ended. Land-based social relations are not automatically changed when official documents are changed: conversely, actually existing land-based social relations may dynamically change while official documents remain the same.

Therefore, in order to address the varying land-based social relations existing in society, multiple land policies have become necessary even in one national setting. These can be in the form of land redistribution, land restitution, tenancy reform, land stewardship and so on. Formal land ownership that is the subject of reform can relate to land controlled by the state, community or private entity. The organization of reformed access to or control over land resources can be by individuals, groups, communities or states. The bottom line is about reforming land-based social relations. These relations encompass the reform of the terms under which land-based wealth is created, appropriated, disposed and consumed, as well as the ways and means by which such processes are effectively controlled by different groups, which entails power relations.

Dynamics of Reform

The way state land laws and land policies are actually implemented results in policy processes and outcomes that affect the pre-existing land-based social relations, which can be broadly categorized as either pro-poor or anti-poor. They are rarely neutral. It is critical to be conscious of the broad trajectories and outcomes of land policies' impact on existing land-based social relations. There are at least four possible broad trajectories, as summarized in Table 8.1.

Redistribution

The first possible trajectory is redistribution, the defining principle for which is that the land-based wealth and power are transferred from the monopoly control of either private landed classes or the state to landless and near-landless working poor (poor peasants and rural labourers). It changes the relative shares of land-based wealth of groups in society. It is a zero-sum reform process (Fox 1993: 10).

Redistributed wealth and power is a matter of degree, depending on the net loss of the landed entities and on the net gain of the landless and near-landless poor. Policies that expropriate lands without compensation and distribute these to peasants are redistributive reforms. A policy that acquires land at usually slightly below the commercial market value and re-sells it to peasants at same price is also redistributive. This is the more common type of land reform, where the importance of land (and so, land redistribution) is largely based on economic or monetary value. Viewing from a redistributive land reform lens provides a useful comparative perspective on contemporary land reform policies. We can for example, determine which national land reform policies (in, say, Brazil, the Philippines and South Africa) are truly redistributive, or which components of a national land reform policy are redistributive in nature and which are not. In addition, having a clear definition of what a redistributive land reform is provides a useful analytic framework for understanding the broad contexts within which rural social movements and state actors interact with each other to facilitate or obstruct redistributive reform. The monetary value-centred way of measuring redistribution (i.e., whether the latter was "confiscation without compensation," "acquisition with some compensation" or "acquisition through full commercial value") is admittedly an important but inherently limited way of measuring the degree of redistributed wealth and power. To many people, land has a value or values that cannot be reduced to or expressed in any monetary value because these are cultural, religious, environmental, social or political.

Just the same, a clearer measure of the land-based wealth and power redistribution can be captured in the concept of redistribution being a matter of degree (Borras 2007, chapter 2). The key is to be able to establish the degree and direction of redistributed wealth and power. For example, an area may be officially classified as idle state land but is actually crop land controlled by a private elite, as in the case of many of the rapidly expanding oil palm plantations in Indonesia. A policy that takes this land away from the controlling elite and redistributes it to landless and near-landless rural poor or indigenous communities will result in real redistributive reform.

Table 8.1: Trajectories of Change and Reform in Land Policies

Type of Reform	Dynamics of Change & Reform; Flow of Wealth & Power Transfers	Remarks
Redistribution	Land-based wealth & power transfers from landed classes or state or community to landless or near-landless working poor	Reform can occur in private or public lands, can involve transfer of full ownership or not, can be received individually or by group
Distribution	Land-based wealth & power received by landless or near-landless working poor without any landed classes losing in the process; state transfers	Reform usually occurs in public lands, can involve transfer of right to alienate or not, can be received individually or by group
Non-(re)distribution	Land-based wealth & power remain in the hands of the few landed classes or the state or community, i.e., status quo that is exclusionary	"No land policy is a policy"; also included are land policies that formalize the exclusionary land claims/rights of landed classes or non-poor elites, including the state or community groups
(Re)concentration	Land-based wealth & power transfers from the state, community or small family farm holders to landed classes, corporate entities, state or community groups	Change dynamics can occur in private or public lands, can involve full transfer of full ownership or not, can be received individually, by group or by corporate entity

Distribution

The second type of reform is distribution. As in redistribution, the landless and near-landless working poor are the recipients of land-based wealth and power. However, the original source of wealth and power can either be the state or community, or a private entity that has been fully compensated by the state. In many settings, this type of reform means affirming and protecting pre-existing land access and occupancy by poor peasants whose tenure is "insecure," partly because of lack of formally recognized property rights. It is a "positive sum" reform process in that it does not take resources from one group in society to redistribute to another. In fact, often such reforms are passed precisely to avoid having to resort to redistributive policies (Fox 1993: 10). For example, a piece of land that is officially categorized as public or state forest is actually an agro-forest tended and tilled by poor peasants or forest dwellers, as in particular cases in the Philippines examined in Borras (2007). A long-term forest-land-use-rights allocation is issued to the poor peasants or forest dwellers in order to make their pre-existing access to the forest land more formal and secure. This is a distributive reform (Franco 2009).

In other cases, a government may purchase, at market price, a piece of private land and then distribute this to the landless for free or a minimal cost. This type of transaction can, under certain conditions (i.e., when there is a net transfer of wealth and power in favour of landless and near-landless peasants), qualify as distributive reform. Some past public land resettlement programs, such as those in Zimbabwe in the 1980s, fall into this category (but see Scott 1998: 269).

Similar to the discussion under the redistributive type of reform, the landed property rights that are distributed can be private, state or community owned. The forms of organizations receiving distributed landed property rights can be individual, group or cooperative. The distributive type of reform, in general, is perhaps not as controversial or conflictual as the redistributive type because it avoids taking lands from the landed classes. But it is certainly not the case that all such reforms are conflict free. A few successful forest land-allocation experiences and widespread (re)allocation of agricultural lands in Vietnam beginning in the 1990s, where land-based wealth was actually transferred to the poor, were accompanied by significant conflict.

Non-(re)distribution

The third category is non-(re)distribution, the defining nature of which is the maintenance of the status quo, marked by inequity and exclusion in land-based social relations. Here, the most typical land policy is "no land policy." Where there are vast land-based inequities and exclusion, a "no land policy" policy effectively advocates for non-redistribution of land-based wealth and power. A similar effect is created by having a land policy, even a potentially pro-poor land policy, but then keeping this dormant. Other examples include forced evictions carried out by landlords, agri-business and real estate companies in potentially or actually contested landholdings to avoid any forms of land and labour reforms. The post-Apartheid farm dweller evictions in South Africa are an example; plantation workers were evicted from their farms by the owners for various reasons including to evade land redistribution (Wegerif et al. 2005).

(Re)concentration

The fourth type is (re)concentration. In this process, while land-based wealth and power transfers do occur, access to and control over the land resource actually gets (re)concentrated in the hands of the well off. This kind of change can occur in private or public lands. The organization of control over land resources can be through individual, corporate, state or community group institutional arrangements in property rights. The transfer may involve full land ownership or not. Different variations are possible, but the bottom line is the same: the recipients of land-based wealth and power transfers are landed classes, agri-business and other non-poor entities or the state. For example, through the *Basic Law* of 1960 in Indonesia, which is a mandate to carry out national land reform, some peasants did get some access to state lands. However, the same law was used to allocate more lands to private corporations engaged in large-scale plantations than to poor peasants. In another example, through the current land policy of Egypt, lands have been taken back from land reform beneficiaries and redistributed to former landlords and other elite entities (Bush 2007).

Struggles around Land Policies & Food Sovereignty

The two last trajectories in land policies, namely non-(re)distribution and (re)concentration, are policies that fundamentally conflict with the basic principles of food sovereignty. Meanwhile, the first two land policy trajectories, redistribution and distribution, are reforms in land-based social relations that are requisites for food sovereignty. The key problem is that today, worldwide, the non-(re)redistribution and (re)concentration land policy types are clearly dominant and seem to be gaining even more ground. Ironically, these policies are almost always promoted as being "pro-poor," for example, in the World Bank 2003 land policy report and policy framework (World Bank 2003). Meanwhile the (re)distributive land policy types seem to be getting increasingly marginalized in mainstream land policy arenas. Given the multiple types of land policies and their actual and potential impact upon existing land-based social relations, it is important to use the typology offered here based on the trajectory of land-based wealth and power transfers.

Discussions about the relationship between agrarian reform and food sovereignty in activist circles seem to privilege the conventional notion of agrarian reform, which is redistributing large landholdings from landed classes to landless peasants and rural labourers in order to create small family farmers. This notion of land reform is of course very important and relevant to a large section of the world's rural poor. Nevertheless, it misses a significant section of the rural poor, those who do not have problems regarding large private landholdings controlled by landed classes. For instance, in many land abundant countries in Africa, key issues include competing claims among community members over land, conflict over grazing areas among pastorialists and between pastorialists and sedentary farmers, actual and potential threat of dispossession due to widespread formalization and privatization of land rights, among others. Non-conventional agrarian reform framing on land is slowly being addressed, especially with the increasing influence of African agrarian movements in La Vía Campesina (see, e.g., La Vía Campesina 2008). But the discussions tend to remain very general and still miss a significant section of the rural world, especially those settings where transnational agrarian movements such as La Vía Campesina are not present, as in the case of China.

To move the discussion beyond our claim of the positive relationship between (re)distributive agrarian reform and food sovereignty, we need to pursue critical discussions in a disaggregated manner based on the type of land setting. There are three broad terrains of struggle in this context: to promote food sovereignty in politically consolidated lands; secure (re)distributive land policies; and fight against non-(re)distributive and land (re)concentration policies. Such discussions will provide a more nuanced analysis of the land–food sovereignty linkages — and agrarian movements' struggles — within and between countries.

Struggling for Food Sovereignty in Politically Consolidated Lands

During the past few decades, some agrarian communities have secured political control over their land resources through successful land reforms, even though partial on most occasions. For example, between the late 1970s and recently, the Left Front government of West Bengal carried out a leasehold reform benefiting around 1.4 million tenant-households. In Mexico, by 1970, 42.9 percent of agricultural land had been redistributed to 43.4 percent of agricultural households. In El Salvador, from 1980s to the 1990s, 20 percent of agricultural land was redistributed to 12 percent of agricultural households (Borras 2008b: 10). In other examples, rural poor communities have emerged victorious in their struggle for land reform through a variety of state–social movement interactions. With different contexts and in recent years, this includes the MST agrarian reform settlements in Brazil beginning in the 1980s, which settled 18.5 percent of agricultural households in 7.6 percent of the total agricultural land by 2005.

In this type of situation, the key challenge for national and transnational agrarian movements is to promote the idea of food sovereignty as a viable alternative vision. Experiments and initiatives around food sovereignty can take different forms and scale from one setting to the next. These experiments can include a straight de-linking from large-scale industrial agriculture and shift to small family farm endeavours, shifting orientation of crop cultivation by small family farms, linking rural producers with urban consumers or a combination of these. The goal is to make redistributed lands productive in socially and environmentally sustainable ways. In agrarian reform scholarship, a common saying is that while land redistribution is the "heart" of agrarian reform, post-land (re)distribution support service packages and favourable rural development policies are the "soul." The two are inseparable. It is in this context that one should consider the relationship between land reform and food sovereignty.

There are two other equally pressing challenges. The first is how to scale up scattered, localized experiments in food sovereignty. This will eventually entail engagement with the central state, especially if policies have to be reformed and programs have to be carried out society-wide. This is perhaps the most daunting challenge for many members of La Vía Campesina who are politically and organizationally strong and have a sufficient mass base with control over large tracts of land. For example, from 1988 to 2008, the Philippine government redistributed around six million hectares of land to around three million peasant households. This constitutes about half of the agricultural land in the country and about two-fifths of the agricultural households, a significant land redistribution. However, only a very small portion of the land reform beneficiaries received direct support service packages from the central government. Aggravating this situation is the fact that the macro-economic policies of export orientation, privatization and deregulation — resulting in cheap imported agricultural products and high prices for agricultural inputs — have rendered many small-scale family farms less competitive. This has created financial difficulties for land reform beneficiaries, many of whom have already leased out or sold their lands to their former landlords, real estate companies and other corporate interests. How then can land reform-based local experiments in food sovereignty progress or be made viable? One thing is certain: in local experiments in land reform–based food sovereignty initiatives, engaging with policy reforms at the national and global levels remains necessary and urgent (Borras 2007).

The second challenge is how to bring the notion of food sovereignty to settings where the rural poor have relatively secure control over land resources but do not have any direct engagement with transnational agrarian movements (TAMs), such as La Vía Campesina, that promote food sovereignty. For instance, many African countries do not have the landlessness problem that exists in Latin America, and agrarian movements that are in the ideological and political mould of La Vía Campesina are not commonly found in Africa. The same condition can be said to exist in China and in some other regions of the world, such as the former U.S.S.R., Eastern Europe, and the Middle East and Northern Africa (MENA) region. We return to this issue later.

In short, in settings where the rural poor have relatively secure control over land resources, it is vital to strengthen vertical and horizontal consolidation and expansion of food sovereignty experiments and advocacy. This will require creative linking between transnational agrarian movements, their allies and diverse groups of rural actors worldwide.

There are various complications and challenges in this regard. For example, while La Vía Campesina and its allies might not be very keen on market-oriented, capitalist arrangements in land property relations, preferring to pursue broadly socialist, community-based or collective types of land property arrangements, many ex-socialist or "transition" countries may not have the same interest. In China and Vietnam, for example, ordinary peasants want to privatize land property relations and push for less state regulation; they generally dislike collective forms of farming. Essentially, broadening the reach of political networks and the mass base of radical transnational agrarian movements such as La Vía Campesina in the context of land property relations and food sovereignty may run the risk of complicating (and diluting?) the global movement's ideological and political positions on certain fundamental issues.

Struggling for (Re)Distributive Land and Agrarian Policies

In many rural places today, most rural poor do not have access to and/or control over land resources, which are usually under the control of landed classes, the state or the community, which in turn facilitates elite control over these lands. This is the case of existing *latifundia*, the big private landholdings such as those in Brazil and the Philippines and large commercial farms with direct provenance from colonial times, such as those in South Africa. There are also productive agricultural lands that are formally owned by the state but are under the effective control of private elites, such as those in Indonesia.

La Vía Campesina and its affiliates have significant presence in many settings marked by extreme inequality in land access and land ownership between poor peasants and landed elites. Therefore, La Vía Campesina has identified agrarian reform as a key issue for struggle. It launched the global campaign for agrarian reform (GCAR) in 1999, an initiative joined by two global allies, the Foodfirst Information and Action Network and the Land Research and Action Network. The main framework of GCAR is based largely on the Latin American campaign of redistributing *latifundia* to landless rural people. With this framework, the GCAR has become a key initiative to expose and oppose the neoliberal land reform model of "market-led agrarian reform" (MLAR), a land reform policy framework based on the "willing seller–willing buyer" principle promoted by the World Bank. This means lands can be transferred from private large landowners to any willing buyers if both parties agree on a market price. It is a purely voluntary real estate transaction, in contrast to the compulsory nature of conventional state-led land reforms. The GCAR has successfully pushed the World Bank into a politically defensive position on this issue, although the campaign has failed to decisively stop MLAR programs, especially the one in Brazil. But while the GCAR has been successful in some aspects, it has missed important ground in the broader struggle for redistributive and distributive land policies (see Borras 2008a; Borras and Franco 2009).

So far, the GCAR has focused on redistributive land policies — but only partially. Some redistributive land policies have never been highlighted in the campaign, nor have distributive land policies such as tenure reforms and stewardship arrangements. In many African and Asian settings, these types of redistributive and distributive reforms have vast potential to effect real transfers of wealth and power in favour of the rural poor. In Indonesia, for example, the struggle for land is not against *latifundia* owners. Instead, the adversaries are the central government and the corporate entities favoured by the government in terms of land and forest allocation. Hence, the general call "to redistribute *latifundia* type private lands to landless peasants" and the administrative and legal complexity that come with such a call are not relevant in settings where most lands are in non-private property categories, such as in Indonesia (Peluso et al. 2008). The complication of the contemporary Bolivian land disputes also involve public lands that are not easily captured in the conventional political discourses on land reform (Kay and Urioste 2007).

Meanwhile, there are some contexts where agrarian reform also means struggle for labour reforms, either as stand alone or complementary demands. Landless rural labourers are usually the poorest of the poor worldwide. They are also usually the most food-deficit households. In many settings, poor peasants who received lands under land reform programs have since leased these out to companies producing high value crops. Hence, included in the landholders-cum-labourers' most urgent interests are labour reforms (e.g., wage and non-wage benefits, safety at the workplace, right to autonomous union work and so on). Experiments in and calls for food sovereignty that do not take into serious consideration the distinct class interests of the rural labourers are bound to be fundamentally flawed.

In short, it is imperative to broaden the definition of (re)distributive land and agrarian policies and pursue multiple-tracked campaigns to achieve these in many peasant societies. This will be an initial but critical step towards a more vibrant, widespread and effective struggle to build food sovereignty.

Struggling against Non-(Re)distributive and (Re)concentration Policies

As explained earlier, many of today's land policies are labelled "pro poor" when they are in fact "pro elite" and often "anti poor." Often referred to as "land distribution" policies, the actual direction of the distribution flow is away from the poor and towards the elite. Policies for non-(re)distribution can be seen through the widespread drives for privatization and formalization of land property rights that are, in many settings, simply formalizing existing inequality, restitution without redistribution and other forms of counter reforms.

Agrarian movements worldwide are engaged in defensive actions against the following non-(re)distributive and (re)concentration policies: privatization and formalization of land property rights in Mozambique; colonization of the Amazon to scatter land claims makers from other politically explosive states of Brazil; full-scale private land sales in community-held agrarian reform lands in Mexico; reversals of earlier land distribution gains in the Philippines; inroads by tourism companies in Indonesia; displacement and dispossession in the Egyptian countryside; peasant-displacing extractive industries such as mining in many rural areas, such as Guatemala and Turkey; large-scale development projects such as dams in India; and so on.

The land grabs in many developing countries are threatening to dispossess or have actually dispossessed rural dwellers. Foreign companies and foreign governments continue to seize large tracts of lands and convert land use from subsistence crops or agro-forestry to food production for export (for the "food security" of richer countries) or for biofuel production for export (GRAIN 2008). (See chapters 5 and 6 in this volume for analysis of landgrabbing and biofuel production). Commercial pressure on land in developing countries continues to increase. In situations like these, any meaningful initiatives for and experiments in food sovereignty are unlikely to gain ground because the rural poor do not have access to and control over lands and are even being displaced from their communities.

A final key challenge for trans/national agrarian movements is to coordinate and systematize these defensive land struggles worldwide as a distinct front of struggle and frame this within an explicitly food sovereignty perspective for mutually reinforcing effect. One difficulty in this terrain is that many of the current hotspots are places where politically important trans/national agrarian movements do not have significant presence, such as Africa (for the land grab) and China (for the continuing land use change from food production to commercial and industrial purposes). The challenge is to figure out how to bring the notion of food sovereignty to settings where there are no formal movements. These are places that have vibrant "everyday forms of peasant resistance" (e.g., pilferage, foot-dragging, food warehouse raids, arson and so on) to the neoliberal onslaught, but such resistance does not take an organizational form involving formal structures and leadership (Kerkvliet 2009). Usually, resistance takes the form of covert political actions, although this is increasingly everyday large-scale overt peasant resistance, such as in rural China, where villagers openly defy state officials on issues around land reclassification, reallocation and dispossession (O'Brien and Li 2006). As Le Mons Walker (2008) and Malseed (2008) argue, in the context of rural China and Burma, respectively, there are many parallels between the recently emerged transnational agrarian movements (TAMs) such as La Vía Campesina and everyday peasant resistance against neoliberal policies offering great potential for forging solidarity linkages. This is critical because, while La Vía Campesina is a global movement with a very significant mass base, much of the rural world remains outside its organizational and political influence.

Thus, it is important to push the discussion of land policy reforms a little further, in order to understand better the precise nature of the linkages between land reform and food sovereignty. Land policies need to be analyzed in terms of the nature and direction of transfers of wealth and power. Viewing food sovereignty as an analytical backdrop, we see that there are food sovereignty initiatives that have a better starting point because of relatively secure land rights by the poor, there are settings where the starting point for any food sovereignty campaign is to call for (re)distributive land reform and there are situations where the challenge is to launch defensive struggles around land. By doing so, we are able to identify the broad terrains of political struggle and some key challenges for agrarian movements. Our discussion in this chapter is very preliminary, as we try to uncover more questions than answers. We hope that our initial observations of the linkages between land reform and food sovereignty will lead to more vibrant and nuanced action research and political discussions and more focused political action in the struggle for food sovereignty.

References

Bachriadi, D. 2009. "Land, Rural Social Movements and Democratisation in Indonesia." Amsterdam: Transnational Institute. Available at <http://tni.org/article/land-rural-social-movements-and-democratisation-indonesia>.

Borras, S. Jr. 2007. *Pro-Poor Land Reform: A Critique*. Ottawa: University of Ottawa Press.

_____. 2008a. "La Vía Campesina and its Global Campaign for Agrarian Reform." *Journal of Agrarian Change* 8, 2/3.

_____. 2008b. *Competing Views and Strategies on Agrarian Reform: Volume II: Philippine Perspective*. Manila: Ateneo de Manila University Press; Honolulu: University of Hawaii Press

Borras, S. Jr., and J.C. Franco. 2009. "Transnational Campaign for Land and Citizenship Rights." IDS Working Paper Series. University of Sussex, Brighton, UK.

Borras, S. Jr., C. Kay and E. Lahiff (eds.). 2008. *Market-Led Agrarian Reform: Critical Perspectives on Neoliberal Land Policies and the Rural Poor*. London: Routledge.

Bush, R. 2007. "Mubarak's Legacy for Egypt's Rural Poor: Returning Land to the Landlords." In H. Akram-Lodhi, S. Borras and C. Kay (eds.), *Land, Poverty and Livelihoods in an Era of Neoliberal Globalization: Perspectives from Developing and Transition Countries*. London: Routledge.

De Soto, H. 2000. *The Mystery of Capital: Why Capitalism Triumphs in the West and Fails Everywhere Else*. New York: Basic Books.

Fox, J. 1993. *The Politics of Food in Mexico: State Power and Social Mobilization*. Ithaca, NY: Cornell University Press.

Franco, J. 2008. "Making Land Rights Accessible: Social Movements and Political-Legal Innovation in the Rural Philippines." *Journal of Development Studies* 44, 7.

_____. 2009. "Pro-Poor Policy Reform and Governance in State/Public Lands: A Critical Civil Society Perspective." *Land Reform Bulletin*. Rome: FAO.

GRAIN. 2008. "Seized: The 2008 Landgrab for Food and Financial Security." Available at <grain.org/briefings/?id=212>.

Kay, C., and M. Urioste. 2007. "Bolivia's Unfinished Reform: Rural Poverty and

Development Policies." In H. Akram-Lodhi, S. Borras and C. Kay (eds.), *Land, Poverty and Livelihoods in an Era of Neoliberal Globalization: Perspectives from Developing and Transition Countries.* London: Routledge.

Kerkvliet, B. 2009. "Everyday Politics in Peasant Societies (and Ours)." *Journal of Peasant Studies* 36, 1.

La Vía Campesina. 2008. *Food Sovereignty for Africa: A Challenge at Fingertips.* Maputo.

Le Mons Walker, K. 2008. "From Covert to Overt: Everyday Peasant Politics in China and the Implications for Transnational Agrarian Movements." *Journal of Agrarian Change* 8, 2/3.

Malseed, K. 2008. "Where there is no Movement: Local Resistance and the Potential for Solidarity." *Journal of Agrarian Change* 8, 2/3.

Monsalve, S., La Vía Campesina, FoodFirst Information and Action Network International, P. Rosset, Land Research Action Network (LRAN), S.V. Vázquez, International Indian Treaty Council, J.K. Carino, Cordillera Women's Education and Resource Center (CWERC) Philippines, West African Network of Peasant and Agricultural Producers' Organizations (ROPPA). 2006. "Agrarian Reform in the Context of Food Sovereignty, the Right to Food and Cultural Diversity: Land, Territory and Dignity." A Civil Society paper presented at the International Conference on Agrarian Reform and Rural Development, Porto Alegre, Brazil, March. Available at <icarrd.org/en/icard_doc_down/Issue_Paper5.pdf>.

Nyamu-Musembi, C. 2007. "De Soto and Land Relations in Rural Africa: Breathing Life into Dead Theories about Property Rights." *Third World Quarterly* 28, 8.

O'Brien, K., and L. Li. 2006. *Rightful Resistance in China.* Cambridge: Cambridge University Press.

Peluso, N., S. Affif and N. Fauzi. 2008. "Claiming the Grounds for Reform: Agrarian and Environmental Movements in Indonesia." *Journal of Agrarian Change* 8, 2/3.

Rosset, P. 2006. "Agrarian Reform and Food Sovereignty: Inseparable Parts of an Alternative Framework." Paper presented at the Land, Poverty Social Justice and Development Conference, Institute of Social Studies, The Hague, Netherlands, January.

Rosset, P., R. Patel and M. Courville (eds.). 2006. *Promised Land: Competing Visions of Agrarian Reform.* Berkeley: Food First Books.

Scott, J. 1998. *Seeing Like a State: How Certain Schemes to Improve the Human Condition Have Failed.* New Haven, CT: Yale University Press.

Wegerif, M., B. Russell and I. Grundling. 2005. *Still Searching for Security: The Reality of Farm Dweller Evictions in South Africa.* Polokwane North: Nkuzi Development Association; Johannesburg: Social Surveys.

World Bank. 2003. Land Policies for Growth and Poverty Reduction. Washington, DC: World Bank; Oxford: Oxford University Press. (A World Bank Policy Research Report prepared by Klaus Deininger.)

9

Scaling Up Agroecological Approaches for Food Sovereignty in Latin America

Miguel A. Altieri

Global and national forces are challenging the ability of Latin America to feed itself while also redefining the significance and the role of the agricultural sector, which has historically been of a dual nature. On the one side, there is a monocultural, competitive, export orientation, which makes a significant contribution to national economies while bringing a variety of economic, environmental and social problems. These problems include negative impacts on public health, ecosystem integrity and food quality, as well as disruption of traditional rural livelihoods and accelerating indebtedness among farmers (Uphoff 2002). The regional consequences of monoculture specialization are manyfold, including an array of environmental problems, worsening insect pest infestations and higher disease incidence linked to the high use of agro-chemicals and the simplification and genetic uniformity of modern crops. Moreover, the efficiency of applied inputs is decreasing, and yields in most key crops are leveling off. In some places, yields are actually in decline. Growing industrialization and globalization, with their emphasis on export crops (such as transgenic soybeans exported for cattle feed to countries such as China, Europe, the U.S. and others), and the rapidly increasing demand for biofuel crops (sugarcane, maize, soybean, oil palm, eucalyptus, etc.) are reshaping the region's agriculture and food supply, with yet unknown economic, social and ecological impacts and risks.

On the other hand, Latin America's peasant and small-farm sector is still significant, making up 63 percent of farmland (ECLAC 2009). Despite migration to urban areas, especially by young people, the situation has not changed much since the late 1980s, when the peasant population reached about 75 million people, representing almost two-thirds of Latin America's rural population (Ortega 1986). In Brazil alone, for example, more than 4.3 million traditional family farmers (about 85 percent of farmers) now occupy just 24.3 percent of the agricultural land of the country (IBGE 2009). In Ecuador, 91 percent of the 843,000 farms are smallholdings, and in Peru, smallholdings account for 80 percent of the 1.6 million farms (ECLAC 2009). Many of these peasants still use traditional farming systems, which represent microcosms of community-based agriculture, offering promising models for promoting biodiversity, sustaining yields without agro-chemicals and conserving ecological integrity while reaching food security (Altieri and Koohafkan 2008).

During the last two decades, the concepts of food sovereignty and agroeco-logically based production systems have gained much attention. New approaches and technologies blending modern agricultural science and indigenous knowledge systems, spearheaded by peasant organizations, NGOs and some government and academic institutions, are enhancing food security while conserving natural resources, agro-biodiversity and soil and water conservation throughout hundreds of rural communities in the region.

The science of agroecology — the application of ecological concepts and prin-ciples to the design and management of sustainable agroecosystems — provides a framework to assess the complexity of agroecosystems. The idea of agroecology is to develop a type of agriculture that does not depend on high chemical and energy inputs. The emphasis is on agricultural systems in which ecological interactions and synergisms between biological components provide the mechanisms for the system to sponsor its own soil fertility, productivity and crop protection (Altieri 1995). In addition to providing a scientific basis for sustainable and enhanced productivity, agroecology promotes the capability of local communities to innovate, evaluate and adapt themselves through farmer-to-farmer research and grassroots extension approaches. Technological approaches emphasizing diversity, synergy, recycling and integration, and social processes that value community involvement, point to the fact that human resource development is the cornerstone of any strategy aimed at increasing food production. In short, agroecology can have a significant effect on the region's food sovereignty.

Small Farmers Are Key

In Latin America, the number of peasant farms reached about 16 million by the late 1980s and occupied close to 60.5 million hectares, or 34.5 percent of the cultivated land; the average size of these farms was about 1.8 hectares. The con-tribution of peasant agriculture to the general food supply in the region has been significant (DeGrandi 1996). In the 1980s, peasant farms produced approximately 41 percent of the agricultural output for domestic consumption: 5 percent of the maize, 77 percent of the beans and 61 percent of the potatoes. In Brazil, family farms control about 33 percent of the area sown to maize, 61 percent under beans and 64 percent planted to cassava, producing 84 percent of the total cassava and 67 percent of all beans. In Ecuador, the peasant sector occupies more than 50 percent of the area devoted to food crops such as maize, beans, barley and okra, and in Mexico, peasants occupy at least 70 percent of the area assigned to maize and 60 percent of the area under beans (FAO 2001). In addition to the peasant and family farm sector, there are about 50 million individuals belonging to some seven hundred different indigenous groups who live and utilize the humid tropical regions of the world. About two million of these live in the Amazon and Southern Mexico. In Mexico, as recently as the 1980s, half of the humid tropics was utilized by indigenous communities and *ejidos* (communally managed farms), featuring integrated agriculture-forestry systems with production aimed at subsistence and local-regional markets (Toledo et al. 1985).

Small Farms Are More Productive and Resource Conserving

Although conventional wisdom claims that small family farms are backward and unproductive, research shows that small farms are much more productive than large farms if total output is considered rather than yield from a single crop (Rosset et al. 2006). Integrated farming systems, in which the small-scale farmer produces grains, fruits, vegetables, fodder and animal products, out-produce yields per unit of monoculture crops such as corn on large-scale farms (Funes-Monzote 2008). A large farm may produce more corn per hectare than a small farm in which the corn is grown as part of a polyculture that also includes beans, squash, potato and fodder. Yet, for smallholder polycultures, productivity in terms of harvestable products per unit area is higher than under large-scale sole cropping with the same level of management. Yield advantages can range from 20 to 60 percent, because polycultures reduce losses due to weeds, insects and diseases and make more efficient use of water, light and nutrients (Beets 1982; Funes-Monzote 2008). In Mexico, a 1.73 hectare plot of land has to be planted with maize monoculture to produce as much food as 1 hectare planted with a mixture of maize, squash and beans. In addition, the maize-squash-bean polyculture produces up to 4 tons per hectare of dry matter for plowing into the soil, compared with 2 tons in a maize monoculture. Likewise, in Brazil, polycultures containing 12,500 maize plants per hectare and 150,000 bean plants per hectare exhibited a yield advantage of 28 percent (Gliessman 1998).

In terms of overall output, the diversified farm produces much more food, even if measured in dollars. In the U.S., data show that 2-hectare farms produced $15,104 per hectare and netted about $2,902 per hectare; the largest farms, averaging 15,581 hectares, yielded $249 per hectare and netted about $52 per hectare (Rosset 2006). Not only do small to medium sized farms produce higher yields than large farms, but they do so with much lower negative impact on the environment. In this regard, small farms are multi-functional; they can be more productive, more efficient and contribute more to economic development than can large farms. Despite the fact that a proportion of medium- and small-scale farms are conventional, in many cases small farmers also take better care of natural resources, including reducing soil erosion and conserving biodiversity (Rosset et al. 2006).

The inverse relationship between farm size and output can be attributed to the more efficient use of land, water and other agricultural resources that usually results from the management of biodiverse farms by small farmers (Funes-Monzote 2008). In terms of converting inputs into outputs, then, on a per unit basis, food sovereignty is more likely to be achieved through the work of small-scale farmers. Building strong rural economies in the Global South based on productive small-scale farming will help stem the tide of out-migration and allow people to remain with their families. As population continues to grow and the amount of farmland and water available to each person continues to shrink, a small-farm structure may become central to feeding the planet, especially as large-scale agriculture increasingly devotes itself to feeding car tanks.

Diversified Farms as Models of Sustainability

In Latin America, the persistence of more than three million agricultural hectares under ancient, traditional management in the form of raised fields, terraces, polycultures, agro-forestry systems and so on demonstrates a successful indigenous agricultural strategy and comprises a tribute to the creativity of traditional farmers. These microcosms of traditional agriculture offer promising models for other areas as they promote biodiversity, thrive without agro-chemicals and sustain year-round multicrop yields (Altieri 1999).

One such sustainable traditional system is the *frijol tapado*, used to produce beans in mid-elevation areas of Central America, on steep slopes with high amounts of rainfall, where most beans in the region are grown. To begin the process, farmers choose a fallow field that is two to three years old so that the woody vegetation dominates the grasses. If the fallow period is less than two years, the grasses will crowd out the emerging bean plants and soil fertility will not have been fully restored since the last harvest. Next, paths are cut through the field with machetes. Then, bean seeds are thrown, or broadcasted, into the fallow vegetation. Finally, the fallow vegetation is cut down into a mulch, which decays and provides nutrients to the maturing bean seedlings. Approximately twelve weeks after broadcasting, the beans are harvested. In Costa Rica, an estimated 60 to 70 percent of beans are produced by *frijol tapado*. Compared to the more labour- and chemical-intensive methods of bean production used by some smallholders, the *tapado* system has a higher rate of return because of lower labour and input costs (Buckles et al. 1998). The *tapado* system allows production of beans for both home consumption and cash to supplement meagre incomes. The cost-effective benefits include no need for expensive and potentially toxic chemicals such as fertilizers and pesticides and a relatively low labour requirement. Soil erosion is minimized because the continuous vegetation cover protects the bare ground from heavy rainfall.

The rationale of the *frijol tapado* has led to the use of green manures as an ecological pathway to the intensification of the maize-bean polyculture, or *milpa*, in areas where long fallows are no longer possible due to population growth or conversion of forest to pasture. After the maize is harvested, the field is abandoned to the previously broadcast *mucuna pruriens* (velvetbean, a leguminous cover crop), leaving a thick mulch layer year round. The velvetbean mulch layer results in improved mineral nutrition in the maize crop, cumulative soil fertility and reduced soil erosion (Altieri 2002). Experiences in Central America show that *mucuna*-based maize systems are fairly stable, allowing respectable yield levels (usually 2–4 tons per hectare) every year. In particular, the system appears to greatly diminish drought stress because the mulch layer helps conserve water in the soil profile. Adequate water in the soil allows nutrients to be readily available to the major crop. In addition, the *mucuna* system suppresses weeds, either because velvetbeans physically prevent them from emerging or surviving, or because a shallow rooting of weeds in the litter layer–soil interface makes them easier to control. Data show that this system, grounded in farmers' knowledge and involving the continuous annual rotation of velvetbean and maize, can be sustained for at least fifteen years at a reasonably high level of productivity without any apparent decline in the natural resource base (Flores 1989). As illustrated with the case of the *mucuna* system, an understanding of the agroecology of traditional farming systems can contribute to the development of contemporary systems. This awareness can only result from integrative studies that determine the myriad factors that condition how farmers perceive their environment and subsequently how they modify it.

In addition to mixing crops, many resource-poor farmers also exploit diversity by growing different varieties of the same crop at the same time and in the same field. In a worldwide survey of crop-varietal diversity on farms, involving twenty-seven crops, Jarvis et al. (2007) found that considerable crop genetic diversity continues to be maintained in traditional crop varieties, especially of major staple crops. In most cases, farmers maintain such diversity as insurance to meet future environmental change or social and economic needs. Many researchers conclude that variety richness enhances productivity and reduces yield variability (Brookfield and Padoch 1994).

Undoubtedly, the ensemble of traditional crop-management practices used by many resource-poor farmers represents a rich resource for modern workers seeking to create novel agroecosystems well adapted to the local agroecological and socioeconomic circumstances of peasants. Peasants use a diversity of techniques, many of which fit well to local conditions. The techniques tend to be knowledge intensive rather than input intensive, and many are site specific; if they are applied to other environments, modifications and adaptations may be necessary. It is vital to maintain the foundations of such modifications grounded in peasants' rationale and knowledge.

Small Farms Are More Resilient to Climate Change

In traditional agroecosystems, the prevalence of complex and diverse cropping systems is of key importance to the stability of peasant farming systems, allowing crops to reach acceptable productivity levels even in the midst of environmentally stressful conditions. In general, traditional agroecosystems are less vulnerable to catastrophic loss because a wide variety of crops is grown in various spatial and temporal arrangements. Research suggests that many small farmers cope and even prepare for climate change, minimizing crop failure through increased use of drought-tolerant local varieties, water harvesting, mixed cropping, opportunistic weeding, agro-forestry and a series of other traditional techniques (Altieri and Koohafkan 2008).

Polycultures exhibit greater yield stability and lower productivity declines during a drought than do monocultures. Natarajan and Willey (1986) examined the effect of drought on enhanced yields with polycultures by manipulating water stress on combinations of sorghum, peanut and millet and on mono-crops of peanut, sorghum and millet. All the polycultures yielded consistently more than corresponding monocultures at five levels of moisture availability, ranging from 297 to 584 mm of water applied over the cropping season. Interestingly, the rate of overyielding actually increased with water stress, such that the relative differences in productivity between monocultures and polycultures became more accentuated as stress increased. Polycultures thus exhibited greater yield stability overall and lower productivity declines during a drought.

Many farmers grow crops in agro-forestry designs, where shade tree cover protects crop plants against extremes in micro-climate and soil moisture fluctuation. Farmers also influence micro-climate by retaining and planting trees, which reduce temperature, wind velocity, evaporation and direct exposure to sunlight and intercept hail and rain. Lin (2007) found that in coffee agroecosystems in Chiapas, Mexico, temperature, humidity and solar radiation fluctuations increased significantly as shade cover decreased. She concluded that shade cover was directly related to the mitigation of variability in micro-climate and soil moisture for the coffee crop.

Surveys conducted on hillsides after Hurricane Mitch in Central America showed that farmers using sustainable practices such as *mucuna* cover crops, inter-cropping and agro-forestry suffered less from mudslides than their conventional neighbours. A study spanning 360 communities and twenty-four departments in Nicaragua, Honduras and Guatemala showed that diversified plots had 20 to 40 percent more topsoil, greater soil moisture, less erosion and experienced lower economic losses than neighbours using monocultures (Holt-Gimenez 2001). Thus, a re-evaluation of indigenous technology can serve as a key source of information on adaptive capacity and resilient capabilities of small farms, features of strategic importance for world farmers in the face of climatic change. Indigenous technologies often reflect a worldview and an understanding of our relationship to the natural world that is more realistic and sustainable than those of western European heritage.

Enhancing the Productivity of Small-Farm Systems through Agroecology

Despite the evidence of the resiliency and productivity advantages of small-scale and traditional farming systems, many scientists and development officials argue that the performance of subsistence agriculture is unsatisfactory and that intensification of production is essential for the transition from subsistence to commercial production. While subsistence farming has not generally produced a meaningful marketable surplus due to land and labour constraints, subsistence farming has the potential to ensure food security (Altieri 1999). Many people wrongly believe that traditional systems do not produce more because hand tools and draft animals put a ceiling on productivity. Productivity may be low, but the cause appears to be social, not technical. When the farmer succeeds in providing enough food for subsistence, there is no pressure to innovate or to enhance yields. However, research shows that traditional crop and animal combinations can often be adapted to increase productivity when the agroecological structuring of the farm via crop combinations and/or animal integration is improved and when the use of labour and local resources is efficient (Altieri 2002). This approach contrasts strongly with many modern agricultural development projects, characterized by broad-scale technological recommendations, which have ignored the heterogeneity of traditional agriculture, resulting in an inevitable mismatching between agricultural development and the needs and potentials of local people and localities (Altieri 1995).

The failure of top-down development has become even more alarming as economic change, fueled by capital and market penetration, are leading to an ecological breakdown that is starting to destroy the sustainability of traditional agriculture. After creating resource-conserving systems for centuries, traditional cultures in areas such as Mesoamerica and the Andes are now being undermined by external political and economic forces. Biodiversity is decreasing on farms, soil degradation is accelerating, community and social organization is breaking down, genetic resources are being eroded and traditions are being lost. Under this scenario and given commercial pressures and urban demands, the challenge is how to guide such transitions in a way that yields and income are increased without raising the debt of peasants and further exacerbating environmental degradation. We contend that this can be done by generating and promoting agroecologically based resource-conserving technologies, a source of which are the very traditional systems that global, industrial monocultural farming is destroying.

Ecological Potential of Traditional Systems

As the inability of the Green Revolution to improve production and farm incomes for the very poor became apparent, the new enthusiasm for ancient technologies spearheaded a quest in Latin America for affordable, productive and ecologically sound technologies that enhance small-farm productivity while conserving resources. One of the early projects advocating this agroecological approach occurred in the mid 1970s, when the former Mexican National Research Institute on Biotic Resources (INIREB by its Spanish acronym) unveiled a plan to build *chinampas* in the swampy region of Veracruz and Tabasco. Perfected by the Aztec inhabitants of the Valley of Mexico prior to the Spanish Conquest, *chinampa* agriculture, a self-sustaining system of raised farming beds in shallow lakes or marshes that has operated for centuries, is one of the most intensive and productive ever devised by humans. It demanded no significant capital inputs yet maintained extraordinarily high yields year after year. A wide variety of staple crops, vegetables and flowers were mixed with an array of fruit from small trees and bushes. In addition, abundant aquatic life in the canals provided valuable sources of protein for the local diet (Gliessman 1998).

According to Sanders (1957), in the mid 1950s, *chinampas* exhibited maize yields of 3.5–6.3 tons per hectare. At that time, these were the highest long-term yields achieved anywhere in Mexico. (In comparison, average maize yields in the U.S. in 1955 were 2.6 tons per hectare and did not pass the 4 tons per hectare mark until 1965). Each hectare of *chinampa* could produce enough food for fifteen to twenty persons per year at modern subsistence levels. Later research indicated that each *chinampero* could work about three quarters of a hectare per year (Jimenez-Osornio and del Amo 1986), meaning that each farmer can support twelve to fifteen people.

Threatened by the growth of Mexico City, *chinampas* nearly vanished except in a few isolated areas. Noting that this system offered a promising model for other areas as it promotes biological diversity, thrives without chemical inputs and sustains year-round yields, INIREB began to promote *chinampas* in the lowland tropics of Mexico. Although implementation and adoption of *chinampas* in Tabasco was somewhat successful, one criticism of the project was that no market outlets were explored for the outputs produced by the community. Nonetheless, the raised beds of Tabasco (or *camellones chontales*) are still in full operation in the swamps of this region, under full control of the Chontal Nation. These "swamp farmers" use traditional agriculture, and the new raised beds produce a great variety of products that provide income and food security.

In the totally different ecoregion of the Andes, several institutions have engaged in programs to restore abandoned terraces and build new ones. In the Colca Valley of Southern Peru, the Programa de Acondicionamiento Territorial y Vivienda Rural (Rural Housing and Territorial Development Program) sponsors terrace reconstruction by offering peasant communities low-interest loans, seeds and other inputs to restore large areas of abandoned terraces. The main advantages of terraces are that they minimize risks in times of frost and/or drought, reduce soil loss, amplify the cropping options because of the micro-climate and hydraulic advantages and improve crop yields. Yield data from new bench terraces showed a 43–65 percent yield increase in potatoes, maize and barley compared to yields of these crops grown on sloping fields (Browder 1989). One of the main constraints of this technology is that it is highly labour intensive, requiring about 350–500 worker days per hectare in a given year. Such demands, however, can be buffered when communities organize and share tasks (Altieri 1995).

In Peru, in search of solutions to contemporary problems of high altitude farming, researchers have uncovered remnants of thousands of hectares of "ridged fields." One fascinating farming effort is the revival of an ingenious system of raised fields that evolved on the high plains of the Peruvian Andes about 3,000 years ago. According to archaeological evidence, these *waru-warus*, platforms of soil surrounded by ditches filled with water, were able to produce bumper crops despite floods, droughts and the killing frost common at altitudes of nearly 4000 metres (Denevan 1995). The *waru-waru* combination of raised beds and canals has proven to have important temperature-moderation effects, extending the growing season and leading to higher productivity compared to chemically fertilized, but normally cultivated, pampa soils. In Camjata, the potato fields reached 13 tons per hectare per year in *waru-warus*. In the Huatta district, reconstructed raised fields also produced impressive harvests, exhibiting a sustained potato yield of 8–14 tons per hectare per year (Browder 1989). These figures contrast favourably with the average potato yields of 1–4 tons per hectare per year produced by other small farmers on the Puna. It is estimated that the initial construction, rebuilding every ten years and annual planting, weeding, harvest and maintenance of raised fields require 270 person-days per hectare per year.

On Chiloe Island in Southern Chile, a secondary centre of origin of potatoes, NGO development workers are tapping the ethno-botanical knowledge of female elders of the Huilliche Nation in an effort to slow genetic erosion and recover some of the original native potato germplasm (Altieri 1995). They intend to make it available to contemporary impoverished farmers, desperately in need of locally adapted varieties that can produce without agro-chemicals. After surveying several agroecosystems of Chiloe, technicians collected hundreds of samples of native potatoes still grown by indigenous farmers. With this material and in collaboration with farmers they established community seed banks where more than 120 traditional varieties are grown year after year and are subjected to selection and seed enhancement. In this way, an *in situ* conservation program was initiated involving several farmers from various rural communities, ensuring the active conservation and exchange of varieties among participating farmers. As more farmers became involved, this strategy allowed a continuous supply of seeds of value to resource-poor farmers for subsistence and also provided a repository of vital genetic diversity for future regional crop-improvement programs (Altieri 2002).

Rural Social Movements & Agroecology

The development of sustainable agriculture requires significant structural changes in addition to technological innovation and farmer-to-farmer solidarity. This is impossible without social movements that create the political will among decision-makers to dismantle and transform the institutions and regulations that presently hold back sustainable agricultural development. For this reason, many argue that a more radical transformation of agriculture is needed, one guided by the notion that ecological change in agriculture cannot be promoted without comparable changes in the social, political, cultural and economic arenas that conform and determine agriculture. The organized peasant- and indigenous-based agrarian movements like La Vía Campesina have long contended that peasants and small-scale farmers need land to produce food for their own communities and for their country. For this reason, they have advocated for genuine agrarian reforms to improve access to and control over land, water, agro-biodiversity and so on, which are of central importance for communities to be able to meet growing food demands. La Vía Campesina believes that in order to protect livelihoods, jobs, people's food security and health as well as the environment, food production has to remain in the hands of peasants and small-scale sustainable farmers and cannot be left under the control of large agri-business companies and supermarket chains. Only by changing the export-led, free-trade-based, industrial agriculture model of large farms can the downward spiral of poverty, low wages, rural-urban migration, hunger and environmental degradation be halted (La Vía Campesina 2008; Rosset 2006). Rural social movements embrace the concept of food sovereignty as an alternative to the neoliberal approach, which puts its faith in an inequitable international trade to solve the world's food problem. Instead, food sovereignty focuses on local autonomy, local markets, local production-consumption cycles, energy and technological sovereignty and farmer-to-farmer networks.

Today's peasant movements understand that dismantling the industrial agrifoods complex and restoring local food systems must be accompanied by the construction of technical and material alternatives that suit the needs of small-scale producers and low-income consumers, while acknowledging geographic and cultural diversity. Researchers can help farmers movements reach food sovereignty and sustainable agriculture by documenting succesful agroecological experiences and sharing such alternative agricultural practices among broad sectors of the rural population via farmer-to-farmer networks (Holt-Gimenez 2006).

Outlook & Prospects

There is no question that small farmers in Latin America can produce much of the needed food for rural and urban communities in the midst of climate change and burgeoning energy costs (Uphoff and Altieri 1999; Pretty et al. 2003). The evidence is conclusive: new agroecological approaches and technologies spearheaded by farmers, NGOs and some local governments around the region are already making a sufficient contribution to food security at the household, national and regional levels. A variety of agroecological and participatory approaches in many countries show very positive outcomes even under adverse environmental conditions. Potentials include raising cereal yields from 50 to 200 percent, increasing stability of production through diversification, improving diets and income, contributing to national food security and even to exports, conservation of the natural-resource base and agro-biodiversity (Uphoff and Altieri 1999). As demonstrated above, many studies show that small, diversified farms can produce from two to ten times more per unit area than can large corporate farms (Funes-Monzote 2008).

Whether the potential and spread of thousands of local agroecological innovations is realized depends on several factors and actions. Proposed agroecological strategies have to deliberately target the poor and not only aim to increase production and conserve natural resources but also to create employment and provide access to local inputs and output markets. New strategies must focus on the facilitation of farmer learning to become experts on agroecology and at capturing the opportunities in their diverse environments (Uphoff 2002).

Researchers and rural development practitioners need to translate general ecological principles and natural resource management concepts into practical advice directly relevant to the needs and circumstances of smallholders. A focus on resource-conserving technologies that use labour efficiently and on diversified farming systems based on natural ecosystem processes is essential. This requires a clear understanding of the relationship between biodiversity and agroecosystem function and identifying management practices and designs that enhance the right kind of biodiversity, which in turn contributes to the maintenance and productivity of agroecosystems (Altieri 1995; Gliessman 1998). Any serious attempt at developing sustainable agricultural technologies must bring to bear local knowledge and skills on the research process (Toledo and Solís 2001). Particular emphasis must be given to involving farmers directly in the formulation of the research agenda and on their active participation in the process of technological innovation and dissemination through *campesino a campesino* (farmer-to-farmer) models that focus on sharing experiences and strengthening local research and problem-solving capacities.

Major changes must be made in policies, institutions and research and development to make sure that agroecological alternatives are adopted, made equitably and broadly accessible and multiplied so that their full benefit for sustainable food security can be realized. Existing subsidies and policy incentives for conventional chemical approaches must be dismantled. Corporate control over the food system must also be challenged. Governments and international public organizations must encourage and support effective partnerships between NGOs, local universities and farmer organizations in order to assist and empower poor farmers to achieve food security, income generation and natural resource conservation.

There is also a need to increase rural incomes through interventions other than enhancing yields, such as complementary marketing and processing activities. Therefore, equitable market opportunities should also be developed that emphasize fair trade, local commercialization and distribution schemes, fair prices and other mechanisms that link farmers and consumers more directly and in solidarity. However, simply opening niche markets for peasant produce among the rich in the North exhibits the same problems of any agro-export scheme that does not prioritize food sovereignty, thus perpetuating dependence and hunger. The ultimate challenge is to increase investment and research in agroecology and scale up projects that have already proven successful to thousands of other farmers. This will generate a meaningful impact on the income, food security and environmental wellbeing of the world's population, especially the millions of poor farmers yet untouched by modern agricultural technology.

Agrarian movements must continue to pressure multinational companies and government officials to ensure that all countries achieve food sovereignty by developing their own domestic farm and food policies that respond to the true needs of their farmers and all consumers, especially the poor. The need to rapidly foster sustainable agriculture requires coalitions among farmers, civil-society organizations (including consumers) and research organizations. Moving towards a more socially just, economically viable and environmentally sound agriculture will be the result of the coordinated action of emerging social movements in the rural sector in alliance with civil-society organizations that are committed to supporting the goals of these farmers movements. The expectation is that, through constant political pressure from organized farmers and members of civil-society, politicians will be pushed to develop and launch policies conducive to enhancing food sovereignty, preserving the natural resource base and ensuring social equity and economic viability.

The new research agenda requires institutional realignments and, if it is to be relevant to peasants and the small- and medium-scale farmers, it must be influenced by agroecology, with its emphasis on complex farming systems, labour demanding techniques and use of organic and local resources. This means that technological solutions have to be location specific and much more information intensive rather than capital intensive. In turn this implies using more farmer knowledge but also providing support to farmers to increase their management skills. Importantly, the agroecological process requires participation and enhancement of farmers' ecological literacy about their farms and resources, laying the foundation for empowerment and continuous innovation by rural communities.

Whether the potential and spread of local agroecological innovations is realized depends on investments, policies and attitude changes on the part of researchers and policymakers. "Greening" the Green Revolution will not be sufficient to reduce hunger and poverty and conserve biodiversity. If the root causes of hunger, poverty and inequity are not confronted head-on, tensions between socially equitable development and ecologically sound conservation are bound to accentuate. Organic farming systems that do not challenge the monocultural nature of plantations and that rely on external inputs and expensive foreign certification seals and fair-trade systems destined only for agro-export offer very little to peasants and small farmers, who become dependent on external inputs and foreign and volatile markets. The fine-tuning of the input-substitution approach will do little to move farmers towards the productive redesign of agroecosystems that would move them away from dependence on external inputs.

References

Altieri, M.A. 1995. *Agroecology: The Science of Sustainable Agriculture.* Boulder, CO: Westview Press.

_____. 1999. "Applying Agroecology to Enhance Productivity of Peasant Farming Systems in Latin America." *Environment, Development and Sustainability* 1.

_____. 2002. "Agroecology: The Science of Natural Resource Management for Poor Farmers in Marginal Environments." *Agriculture, Ecosystems and Environment* 93.

Altieri, M.A., and P. Koohafkan. 2008. *Enduring Farms: Climate Change, Smallholders and Traditional Farming Communities.* Environment and Development Series 6. Malaysia: Third World Network.

Beets, W.G. 1982. *Multiple Cropping and Tropical Farming Systems.* Boulder, CO: Westview Press.

Brookfield, H., and C. Padoch. 1994. "Appreciating Agrobiodiversity: A Look at the Dynamism and Diversity of Indigenous Farming Practices." *Environment* 36.

Browder, J.O. l989. *Fragile Lands in Latin America: Strategies for Sustainable Development.* Boulder CO: Westview Press.

Buckles, D., B. Triomphe and G. Sain. 1998. *Cover Crops in Hillside Agriculture: Farmer Innovation with Mucuna.* Ottawa: International Development Research Center.

DeGrandi, J.C. 1996. *El Desarrollo de los Sistemas de Agricultura Campesina en America Latina: Un Analisis de la Influencia del Contexto Socio-Economico.* Rome: Food and Agriculture Organization.

Denevan, W.M. 1995. "Prehistoric Agricultural Methods as Models for Sustainability." *Advanced Plant Pathology* 11.

ECLAC (Economic Commission for Latin America and the Caribbean). 2009. The Outlook for Agriculture and Rural Development in the Americas: A Perspective on Latin America and the Caribbean. Santiago, Chile: ECLAC-IICA-FAO.

Flores, M. 1989. "Velvetbeans: An Alternative to Improve Small Farmers' Agriculture." *ILEIA Newsletter* 5.

FAO (Food and Agriculture Organization of the United Nations). 2001. FAOSTAT: FAO Statistical Databases. Available at <apps.fao.org>.

Funes-Monzote, F.R. 2008. "Farming Like We're Here to Stay: The Mixed Farming Alternative for Cuba." Ph.D. thesis, Wageningen University, Netherlands.

Gliessman, S.R. 1998. *Agro-ecology: Ecological Process in Sustainable Agriculture.* Michigan: Ann Arbor Press.

Holt-Gimenez, E. 2001. "Measuring Farms' Agroecological Resistance to Hurricane Mitch." *LEISA* 17.

_____. 2006. *Campesino a Campesino: Voices from Latin America's Farmer to Farmer Movement for Sustainable Agriculture.* Oakland: Food First Books.

IBGE (Instituto Brasileiro de Geografia e Estadistica). 2009. "Censo Agropecuario 2006" Available at <ibge.gov.br/home/estatistica/economia/agropecuaria/censoagro/agri_familiar_2006/default.shtm>.

Jimenez-Osornio, J., and S. del Amo. 1986. "An Intensive Mexican Traditional Agro-ecosystem: The Chinampa." Proceedings of 6th International Scientific Conference IFOAM. Santa Cruz, CA.

Jarvis. D.I., C. Padoch and H.D. Cooper. 2007. *Managing Biodiversity in Agricultural Ecosystems.* New York: Columbia University Press.

La Vía Campesina. 2008. "An Answer to the Global Food Crisis: Peasants and Small Farmers Can Feed the World!" Available at <viacampesina.org>.

Lin, B.B. 2007. "Agro-forestry Management as an Adaptive Strategy Against Potential Microclimate Extremes in Coffee Agriculture." *Agricultural and Forest Meteorology* 144.

Natarajan, M., and R.W. Willey. 1986. "The Effects of Water Stress on Yield Advantages of Intercropping Systems." *Field Crops Research* 13.

Ortega, E. 1986. *Peasant Agriculture in Latin America.* Santiago: Joint ECLAC/FAO Agriculture Division.

Pretty, J., J.I.L. Morrison and R.E. Hine. 2003. "Reducing Food Poverty by Increasing Agricultural Sustainability in Developing Countries." *Agriculture, Ecosystems and Environment* 95.

Rosset, P.M. 2006. *Food is Different: Why We Must Get the WTO Out of Agriculture.* Black Point, NS: Fernwood Publishing.

Rosset, P.M., R. Patel and M. Courville. 2006. *Promised Land: Competing Visions of Agrarian Reform.* Oakland, CA: Food First Books.

Sanders, W.T. 1957. "Tierra y Agua: A Study of the Ecological Factors in the Development of Meso-American Civilizations." PhD dissertation, Harvard University.

Toledo, V.M., J. Carabias, C. Mapes and C. Toledo. 1985. *Ecologia y Autosuficiencia Alimentaria.* Mexico City: Siglo Veintiuno Editores.

Toledo, V.M., and L. Solís. 2001. "Ciencia para los Pobres: El Programa 'Agua para Siempre' de la Regiòn Mixteca." *Ciencias* 64.

Uphoff, N. 2002. *Agroecological Innovations: Increasing Food Production with Participatory Development.* London: Earthscan.

Uphoff, N., and M.A. Altieri. 1999. *Alternatives to Conventional Modern Agriculture for Meeting World Food Needs in the Next Century.* Ithaca, NY: Cornell International Institute for Food, Agriculture and Development.

10

Unearthing the Cultural & Material Struggles over Seed in Malawi

Rachel Bezner Kerr

A peasant farmer plants a seed, which grows into a plant, and that plant is harvested and eaten. These fundamental human and biological activities govern, in very simple terms, human survival. Because of the critical role that seeds play, not only in providing food but also in sustaining cultural knowledge and protecting agro-biodiversity, La Vía Campesina (2007) has made seeds and seed sovereignty a focal point of resistance and action. Seeds have also been the site of increased focus for agri-business, where multinational corporations, in concert with the international agricultural research community, have gained considerable control over seed resources in the last four decades (Kloppenburg 2004).

Many peasant farmers have limited control over what type of seeds to grow, how to plant them and whether they can keep, share or sell seeds. Thus, the social dynamics surrounding seeds are an important element in struggles for food sovereignty between men and women, different generations, communities, the state, scientists and private corporations. This chapter examines these social dynamics from the perspective of farmers living in the 500 square kilometre region surrounding the town of Ekwendeni in Northern Malawi, in South-Central Africa.[1]

Food sovereignty is an issue of vital importance to most Malawian farmers. The majority of Malawians are peasant farmers, growing maize as their staple crop and groundnuts and other crops for sale and consumption and relying on their own production for seed stock. In the last decade, between 70 and 85 percent of households in Malawi ran out of their own food stocks months prior to the next harvest and coped by reducing meals, eating seeds set aside for planting and doing *ganyu*, informal, short-term farm labour done for food, cash or seeds (Bezner Kerr 2005b). As a result, the average child in Malawi does not have access to adequate food, while almost half of children under five years old showed indications of chronic under-nutrition for the past two decades (NSO and ORC Macro 2001, 2005). A major cause and effect of food insecurity in Malawi is lack of access to seeds (Devereux 1997).

A Brief History of Seeds & Agriculture in Malawi

Erosion of seed sovereignty in Malawi has been a historical process, beginning with the British colonial promotion of particular farming practices, crop types and varieties. During the British colonial period (1889–1964), missions, settlers and foreign companies took over 15 percent of the most fertile land (Davison 1997). Malawi had fewer European settlers than other British colonies in the region because of a lack of significant mineral wealth. As a result, colonial policy supported African farming more so than in other colonies (Kydd and Christiansen 1982); however, the emphasis was on export production for the British Empire (McCracken 1982). A "hut" tax required many Malawian men to migrate to Southern Rhodesia or South Africa to work in mines or estates, undermining African peasant efforts to maintain their livelihoods. Women became increasingly responsible for household farming, which deepened gender inequalities in the division of labour.

After World War II came an increased colonial focus on "modernizing" African agriculture by increasing yields and freeing up land for cash crop production (Kalinga 1993). State support in the form of agricultural extension, credit and taxes supported wealthier African farmers at the expense of poorer farmers (McCracken 1982). The colonial government supported a distinct group of African large landholders by appointing them to marketing boards and providing them with extension advice, seeds, credit and subsidized fertilizer, while at the same time taxing smallholder farmers considerably (McCracken 1982; Kalinga 1993).

With little or no proof, most colonial agricultural scientists assumed that African local maize seed varieties were of lower quality, with less pest resistance than imported maize seeds from the U.S. and South Africa. African peasant farmers, it was argued by Europeans, needed assistance to learn proper farming techniques to feed themselves and conserve the land. Starting in the second year of its existence, the Nyasaland Department of Agriculture began to experiment with different U.S. and South African dent maize varieties and did selection work to "improve" native seed quality. After just one year of selection work on native varieties (normally plant selection must be carried out over several growing seasons for the results to have any validity), the department agriculturalist abandoned further testing, reporting that the imported varieties were much more resistant to pests than the native varieties: "Under identical conditions the selected native varieties proved decidedly inferior, and it is proposed to abandon direct selection of this type, restricting experiments to the infusion of new vigour by hybridisation with the better imported varieties" (Davy 1912: 17).

Following independence in 1964, postcolonial policies under a repressive one-party state led by Kamuzu Banda continued most of the colonial policies, emphasizing export crops, discouraging cultivation of diversified peasant crops such as finger millet and denigrating local peasant agricultural practice by stressing modernization through the use of fertilizer and hybrid seeds (McCann 2005). Smallholder farmers were required to sell their maize and tobacco crops through national marketing boards at low prices while estates could sell their crops directly to multinational companies. The surplus money earned by marketing boards was primarily directed towards banks and the estate sector rather than returned to smallholder farmers (Kydd and Christiansen 1982).

In the late 1970s, a combination of rising oil prices, falling tobacco prices, drought and the loss of transportation networks through Mozambique due to the civil war led to a debt crisis (Lele 1989). Structural adjustment policies imposed by the World Bank and IMF in 1981 and market liberalization in 1994 involved the removal of fertilizer subsidies, devaluation of the Malawian currency and reduction of government services such as extension (Harrigan 2008; Peters 1996). Women peasant farmers, already largely ignored by the agricultural extension system, were most affected by the structural adjustment programs through both reductions in health care services, which increased their workload, and reduced access to fertilizer and seed (Gladwin 1992).

Of most importance in terms of affecting access to seed was the privatization of the National Seed Company of Malawi (NSCM). The NSCM, a parastatal organization formed during the postcolonial era to produce seed varieties bred by the Ministry of Agriculture, was sold to Cargill, and in 1996 Cargill sold it to Monsanto. The NSCM had produced legume seeds (e.g., beans, groundnuts), which were then sold in rural agricultural depots. But, with privatization, all legumes and other self-pollinating plants were removed from production, and hybrid maize became the primary focus of NSCM.

National funding also declined for agricultural research and extension, which became increasingly reliant on outside donors, particularly agri-business.[2] Despite a successful transition to multi-party democracy in 1993, governments continued to follow neoliberal economic reforms and implemented market liberalization, leading to heightened inequality and increased power and influence in rural areas for foreign governments, multilateral institutions, NGOs, corporations and international agricultural research institutions. Their interaction with the government and local communities are important in determining peasant access to and control over seeds.

Seeds as Multiple Sites of Struggles over Sovereignty

Smallholder farmers in Malawi's Ekwendeni region grow, select, share, eat and sell maize and groundnut seed. Maize has been an important commodity for African peasant farmers in Malawi since at least the late 1800s (Smale 1995). Reliance on maize as a cash source increased with the implementation of structural adjustment policies, and maize is currently a critical source of cash for food insecure farmers (Muhr et al. 1999). Groundnut is the most common legume in Malawi both in terms of total production and area under cultivation, both of which have risen in the last decade (FAOSTAT 2010; Freeman et al. 2002), and African peasant farmers have exported groundnuts for more than a century, at least in part due to colonial policy (Peters 1996). Over the last two decades there has been an increased demand for legumes in general, particularly pigeon peas but also groundnuts, by urban traders in Malawi (FAOSTAT 2010; Muhr et al. 1999).

Both maize and groundnut crops have been in Malawi for several hundred years, and each crop has a particular "story" rich with political, economic and social struggle and meaning. Power relations, knowledge sharing and development are embedded in different ways in struggles over maize and groundnut seeds in Northern Malawi. I use the biological aspects of the seed to draw out these relationships. The six important characteristics of maize and groundnut seed shown in Table 10.1 are intimately linked to social practices and relations and the struggle for seed sovereignty. First of all, the seed is a means of production in that growing seed produces food. Historically, intercropping of groundnuts and maize has

Table 10.1: Comparison of Maize & Groundnut Seeds in Malawi

Seed Characteristics	Maize	Groundnut
Means of Production	Requires considerable nitrogen; adds carbon, organic matter to soil	Adds nitrogen to the soil
Commodity	Important commercial and local crop in Malawi; third largest staple crop in the world	Important export crop for Malawi and important local economic crop
Food	Main food crop in Malawi, makes up to 70–80% of diet (Ferguson and Gibson 1993)	Minor food crop in Malawi. Important source of protein, oil and iron (Hotz 2001)
Gift or Currency	Important crop for seed gifts and exchange. Day labourers and sharecroppers paid in maize	Important crop for seed gifts and exchange. Farmers may work for others to receive seed
Genetic Information	Cross-pollinating; hard to maintain characteristics. Rapid multiplication rate	Self-pollinating; easy to maintain stable varieties. Slow multiplication rate
"Bred" with Cultural Values	Grown since the 16th century in Malawi; is considered "life giving." Women usually select local varieties, but men usually make decisions about hybrids	Grown since the 16th century in Malawi. Considered the "wife of the home" because it supplies oil and seasoning to food. Women often control decisions

Source: Bezner Kerr 2006: 115.

been practised to take advantage of the former's ability to fix nitrogen in the soil. However, efforts to promote "modern" farming often depict intercropping or other traditional agroecological approaches as backward.

Second, the seed has become a commodity, a standardized product (as grain or germplasm) sold on the global market, disembedded from the social relations and environmental conditions that produced it. In Malawi, the implementation of structural adjustment programs involving privatization and agricultural liberalization led to the successful entry of transnational companies into seed production and distribution, thus greatly increasing the role of seed as an international commodity. The relative role that maize and groundnuts play as commodities in Malawi is now intimately connected to the corporate food regime's neoliberal project of increased market control over seeds and crops (Weis 2007). The economic and agricultural policies pursued in Malawi, and in particular the role of Monsanto in promoting particular seed types, have major implications for what kinds of seeds are grown and sold by Malawian farmers, as the discussion on maize seed varieties below illustrates.

Third, seed itself is often food, as in the case of most grains. This issue is particularly pertinent for Malawi, where many households have to make choices about whether to save seed for next year, sell seed for cash or eat the seed (Bezner Kerr 2005a, 2005b). Since maize is the primary food crop in Malawi, making up over 50 percent of a Malawian diet (McCann 2005) and farmers also use it as a primary cash source, many farmers face difficult choices with their maize harvest. Unlike other countries in Southern Africa where sorghum and millet continue to be an important primary staple, in Malawi maize replaced these grains as the primary staple in the twentieth century. Groundnuts have also been a common food source in Malawi for at least a hundred years (Tindall 1968; Berry and Petty 1992), providing an important source of protein, oil and iron in the diet (Hotz and Gibson 2001). This nutritional role makes the contestation of groundnut seed at the household level even more important. NGO efforts to increase peasant groundnut production as a means to improve nutrition and income have been common throughout Malawi in recent years. However, there may be disputes between men and women at the household level about whether to sell groundnuts or feed them to young children to improve their nutrition (Bezner Kerr 2008).

Fourth, seed in Malawi, particularly maize, is used as a gift or a type of currency in exchange for labour. The use to which seed is put is in part seasonally determined. During the planting season, people might work for others in exchange for seed as planting material; during the rainy season, when households are low on food stocks, people might work for others in exchange for seed as food (Devereux 1997). Kin often give seed as gifts when they visit each other or in times of need. Gifts have been identified as an important site where power relations are reproduced or resisted (Bourdieu 1977). Gifts can also be a form of knowledge transfer. In Northern Malawi, women give maize and groundnut seed to their daughters-in-law upon marriage. At this time they often instruct their daughters-in-law about how to store and select seed for future harvest.

Fifth, the seed contains genetic information that can be crossed through plant breeding to produce new varieties. Here the reproductive differences between maize and groundnuts become critical. There are three major categories of maize seed available in Malawi: hybrid, open-pollinated varieties (OPVs) and local maize.

To produce a hybrid maize variety, two or more selected genetic lines are bred together or crossed to produce a progeny population that is often dramatically more productive than either parent, resulting in hybrid vigour. In order to maintain the varietal quality, farmers must purchase new hybrid maize seed annually. Hybrid maize breeding and production also typically require scientific involvement and a fairly controlled breeding environment.[3]

An open-pollinated variety of maize is a genetically heterogeneous population of plants that has been selected for particular traits and for general agronomic uniformity. OPVs can be recycled for several years before they need to be replaced and are often promoted by scientists, who feel that they meet the middle ground between agronomic needs (e.g., yield, drought tolerance) and reproduction at the farm level.

Local maize types, also called folk varieties or landraces, are genetically diverse populations of maize varieties, with variable traits (Brush 2004), which are kept for their cultural significance as well as their agroecological properties. Local maize is also considered to perform better under poor conditions compared to hybrid maize. One peasant farmer stated: "Local maize, we really like it because sometimes even when you don't have fertilizer, if you plant at the end you get something, but if you plant the *boma* [government] type you really don't get anything without fertilizer" (Focus group discussion with fifteen farmers and author, October 9, 2003).

There are also important differences between flint and dent maize varieties in Malawi. Flint maize varieties (which can be local, OPV or hybrid) produce more flour from mortar pounding, which is used by most Malawians to produce refined maize flour for their staple food, *nsima* (Smale and Jayne 2003). Flint maize also stores better in a typical granary in Malawi, being less prone to pest infestation because of the tighter husk cover and harder grains (Rusike and Smale 1998). Dent maize types, which make up the vast majority of hybrid maize types sold in Malawi, have softer starch granules in the kernel compared to flint maize, produce less flour and are more prone to pest infestation.

Reproduction of hybrid varieties is technically difficult for smallholder farmers, whereas open-pollinated and local varieties can easily be reproduced at the farm level. The majority of Malawian peasant farmers grow a mixture of local and hybrid varieties of maize. Due to cross pollination and small land sizes, as well as active efforts by the state and private sector to promote hybrid maize, it is increasingly hard to physically separate local, OPV and hybrid maize, leading to the deterioration and transformation of local maize varieties. The Malawian government's decision to allow genetically modified crops to be tested, as well as the use of GM maize in food distribution during the 2002 famine, also increases the likelihood of contamination of local varieties, as documented in Mexico (Pineyro-Nelson et al. 2009).

Finally, seed in Malawi embodies cultural perceptions, since seed selection practices are fundamentally social practices, which are influenced by ideas about food, agriculture, nature and so on. A narrow focus on seed as a source of food or as a commodity ignores the intimate and multiple links between rural livelihoods, wellbeing and cultural heritage.

Social Practices Linking Maize & Groundnut Seed to Food Sovereignty

There are several key social and biological practices involved with seeds that are related to struggles over seed sovereignty: namely seed exchanges/seed distribution and planting/selection. These practices and struggles ebb and flow with the agricultural seasons. Each agricultural activity has associated social practices through which peasant farmers access seed and from which some groups are excluded.

Obtaining Seed Prior to Planting Season: Exchanges and Distribution

Most smallholder farmers in Malawi obtain their seed through several processes: their own production and from relatives, neighbours and small-scale private traders at the local market. The type of seed determines which source is more important. A survey of three hundred farmers conducted by the author in 2002 found that 35 percent of farmers sourced their local maize seed from on-farm, 16 percent from neighbours or friends and 16 percent from relatives (Bezner Kerr 2006). In contrast, while hybrid maize was obtained from private traders for 45 percent of farmers, 15 percent obtained hybrid maize from on-farm (i.e., recycled hybrid) and 18 percent from relatives. Groundnut seed was more often sourced from private traders (36 percent), followed by neighbours and friends (21 percent) and on-farm (16 percent) (Bezner Kerr 2006).

The state has had a major role in determining seed access through various nationwide distribution activities in Malawi. The Agricultural Development and Marketing Corporation (formerly the Farmers Marketing Board), which previously purchased groundnuts and maize from smallholder farmers, exported it abroad and sold seed, has been semi-privatized and has had its funding severely cut back. Many peasants cannot afford seeds sold by agro-input stores, with maize seed prices increasing tenfold in a decade under structural adjustment (Edriss et al. 2004). Farmers also have reduced access to seed other than maize in the current system. Crops such as sorghum and millet, indigenous grains that are more drought tolerant than maize, are not available through the state system, and hundreds of OPV maize varieties, which have been bred by the Ministry of Agriculture to be more drought and disease resistant, are left "standing on the research shelves" (and thus not available for purchase), as the chief maize breeder quietly bemoaned in an interview (May 15, 2004).

Complicating the changes to the national seed distribution system is the inconsistent and highly politically charged distribution of free or subsidized fertilizer, maize and, at times, legume seed over the past two decades (Harrigan 2008). In the 1990s, the government and NGOs sponsored several small input distribution programs. For example, in 1998 the government began the Starter Pack Program, which involved the free distribution of enough fertilizer, maize and legume seed (e.g., beans, groundnuts, pigeonpeas) for 0.1 hectare[4] (Harrigan 2003, 2008; Levy et al. 2004). After two years the program was scaled down, with free input provision targeted to poor households, and renamed the Targeted Input Program. In 2005–06, with the new government under Bingu wa Mthalika, the Agricultural Input Subsidy Program, a subsidized coupon system, was developed. Currently, the Agricultural Input Subsidy Program is the major engine of seed enterprise in the country, providing large contracts to foreign seed companies, agro-input dealers and farmer cooperatives for seed production and distribution (Dorward et al. 2008b). One coupon allows a farmer to buy sufficient fertilizer for 0.4 hectares at approximately one-third the market value, and over 2.5 million coupons were redeemed (Denning et al. 2009). Various donors, such as the Department for International Development, the European Union and the United Nations Development Fund, funded the input subsidy/free distribution at different times, each with a different vision of its purpose. Some donors see it as a means to target highly food insecure households as a form of "social safety net" that would reduce the need for strategic grain reserves and food aid (Levy et al. 2004). Subsidized fertilizer and hybrid seed are thus widely touted as effective at increasing maize output and reducing food insecurity (Cromwell et al. 2001; Denning et al. 2009; Levy et al. 2004). Those who supported the input subsidy over the years are largely agricultural scientists, donors such as the Rockefeller Foundation, Department for International Development (government aid agency of the United Kingdom) and the E.U., as well as Overseas Development Institute researchers, who evaluated the program (Devereux 2002, Levy et al. 2004).

Others are more cautious about the impacts, suggesting that input subsidies are not a quick fix for food security, that they can become tools of political manipulation, vulnerable to fraud and very expensive or that they undermine the private sector and divert valuable government resources from other agricultural investments (Dorward et al. 2008a; Tripp 2001; Harrigan 2003). The World Bank and USAID have been highly critical of the program, suggesting that it leads to a "maize poverty trap" by fostering dependency on maize (Devereux 2002; Harrigan 2008).

There are also scientific and political debates about what types of seed should be included in the input package. Hybrid maize (i.e., MH18) was distributed in the package for the first two years of the program, followed by OPV maize, and currently farmers can make a choice as to which seed to grow, depending upon availability (Denning et al. 2009). Decision-making about what maize seed type to distribute was negotiated largely by the donors and foreign NGOs (interview with Dr. Sieglinde Snapp, March 18, 2005). For example, Sasakawa Global 2000 entered Malawi in 1998 and began to strongly advocate maize hybrid mono-crops at a higher plant density, without legumes. This NGO had a tremendous amount of resources at its disposal, and the head of the governmental Maize Commodity Team left to work for it. By the third year of the program, the Sasakawa recommendations for maize type and agronomic practices were in the input distribution pack, and legumes had been removed. Some of the agricultural scientists working on the Maize Productivity Task Force were furious at the changes, since they did not reflect scientific recommendations and were not based on scientific evidence from Malawi (Snapp and Blackie 2003).

High uptake of hybrid maize by farmers in the last two years, as a result of the Agriculture Input Subsidy Program's coupon system, has also led to claims that Malawian farmers prefer hybrid maize (Denning et al. 2009). One major evaluation of this program noted that the hybrid seed sector, all four companies of which are multinationals (Monsanto, Pannar, Pioneer and SeedCo[5]), benefited tremendously from the input distribution program:

> Suppliers of hybrid seed began strong promotional campaigns using the voucher program as a means of helping farmers to experiment with hybrids…. Producers and distributors of hybrids felt that the flexible seed voucher had a *very positive impact on their sales* and represented an improvement over earlier government programs that have tended to favor OPVs. (Dorward et al. 2008b: 44, emphasis added)

Since the other three companies producing OPV seed are all small, newly established Malawian companies, the claim that an increase in hybrid maize seed compared to OPV is evidence of farmer preference is questionable. Multinational companies with large advertising budgets, supply chains and capital out-competed these small Malawian companies by reaching more farmers. The issue of OPVs and hybrids is not only a scientific debate about the merits of open-pollinated varieties versus hybrids, it is also about a state- versus market-oriented approach, and the power of donors, corporations and NGOs in seed policies in Malawi. Farmers have little voice in these narratives or practices.

While there has been a dramatic increase in NGO seed distribution, this distribution is patchy and unreliable. Many NGOs are run by elite, urban, middle-class Malawians who are disconnected and at times disdainful of the seed requirements and knowledge of rural Malawian farmers. Many ascribe to the notion that only the better-off farmers should survive, and many also think that hybrid maize and "modern" farming methods are the solution for Malawians' poverty. The agenda of these NGOs is also often set from abroad and may work in concert with foreign governments, international NGOs or the private sector. For example, in Malawi in 2005, the donation of hybrid maize seed by Monsanto, distributed by Sasakawa Global 2000 and the Ministry of Agriculture, highlights the blurry lines between charity and profit that slowly erode seed sovereignty for farmers.

Community and kin sources remain critical for seed exchange and support for those farmers unable to produce enough of their own on-farm seed. Prior to planting season, many people give seeds as gifts, both to kin and friends. Older women play an important role in circulating seed in these exchanges and also in seed selection. The exception for this role is hybrid maize and tobacco, both of which are crops predominantly controlled by men. Older women also provide a wedding gift of an initial start-up seed supply to their daughters-in-law (Bezner Kerr et al. 2008). These generational and gender differences embedded in seed relations are rooted in the structural inequality of women in this patrilineal Tumbuka and Ngoni culture, where men inherit land and women gain status from having sons. An older woman's role as manager in seed selection is built on this patriarchal system. These gender inequalities challenge a notion of food sovereignty rooted in cultural traditions and need to be addressed if seed sovereignty is to foster social equity.

Conversely, seeds circulate within communities as important sources of reciprocal exchange and indigenous knowledge, which challenge neoliberal values of individualism and profit. Seed gifts reinforce important social bonds and are an indirect type of credit, which is often linked to a type of labour exchanged between kin relations called *ulimizgo*. A typical example of this communal form of labour involves one person, often someone with higher status in terms of age or food security, inviting extended family and neighbours to their fields to carry out a particular farming practice, such as harvesting or weeding. In return the kin are given a meal and/or homemade beer. Although the labour practice appears to be rarely directly associated with seed exchange, families that do *ulimizgo* for one another also share seed. Families with limited maize and livestock availability, especially younger married couples, are less likely to hold *ulimizgo,* while older people, especially women, are able to call on kin for this. In turn, people who organize *ulimizgo* give seed to those family members who ask.

Another community source of seeds is that of neighbours and friends in the area. One survey of three hundred farmers indicated that neighbours and friends were the third most common seed source (after private traders and on-farm production), with 65 percent receiving maize or groundnut seed from neighbours or friends (Bezner Kerr 2006: 225). Some peasant farmers also said that they give seed to relatives who live further away because, if their crop fails one year, they might be able to get seed from an area with different rainfall patterns. Thus, giving away seed is also a means of maintaining seed stock in different geographic regions, thereby reinforcing local seed sovereignty and strengthening community resilience in the face of drought or other challenges.

However, better-off farmers in some cases withdraw from community networks of exchange, resulting in reduced seed access for food insecure peasant farmers. Those few better-off households who choose to give seed only to immediate relatives and sell the rest also do not tend to participate in *ulimizgo*, citing it as a waste of time and resources and preferring to hire *ganyu* workers. Interestingly, many of them also lament a reduction in community sharing but attribute this trend to a general lessening in community values and not to their own practices.

Although many food insecure households continue to seek out small amounts of seed from relatives, neighbours and friends, the amount of seed obtained is usually not enough for household nutritional needs. When local villagers ask for seed, people often give seed out of charity and a sense of empathy, arguing that they could not easily refuse a request for seed because "next year it could be me asking for seed" (Interview with smallholder woman farmer, March 4, 2004).

Tenants are largely left out of the community seed sources. While neighbours are willing to give seed to other villagers, they see tenants as temporary residents and do not feel obliged to give them seed. Tenants, increasingly migrating from Southern Malawi due to acute land shortages, are visible examples of dispossessed peasants, displaced by government policy that largely ignores land inequalities (Kanyongolo 2004) and global neoliberal policies, which reinforce these inequalities (Weis 2007). AIDS-affected families are also often unable to source seeds through social exchanges or even obtain *ganyu*, a crucial source of seeds and food during the hungry season. These families are excluded due to social stigma surrounding their disease. One woman noted, "We are viewed as useless and almost ready to die" (Bezner Kerr and Shumba 2007: 9).

While some colonial and postcolonial processes have reinforced and deepened gender and class inequalities, increased corporate control of seeds has at times disrupted the generational control over seeds (Bezner Kerr 2005a). Young people, particularly men, with greater access to capital, have been able to gain greater control over maize seed decision-making but in turn have cut themselves off from seed exchanges, thereby worsening the position of more food insecure families.

Selecting and Planting Seed

While hybrid seed maize types have populated store shelves over time, legume and other open-pollinated crops have become less available on the market or from government sources over the years. Smallholders play a critical role in maintaining seed diversity by selecting and sharing local seed varieties of different open-pollinated crops.

Seed selection can have distinct purposes in relation to genetic status: to retain desired traits of a variety, to adapt a plant to local conditions or to create a new variety. Plants can also be selected due to physiological status, based on the plant vigour or crop health. Where, by whom and how seed selection is carried out has implications for the biological properties of the plant and the people who retain control of that variety. Understanding who does seed selection within the household gives information about who has knowledge and power about what type of seed is important and the implications for seed access and control. For example, older women, usually together with several daughters and daughters-in-law, carry out seed selection after harvest as part of their normal agricultural tasks. The role that older women hold as seed provisioners is linked to their broader role in decision-making and control over child care (Bezner Kerr et al. 2008).

Beyond the household, what type of maize seed to plant has been a major site of struggle over seed sovereignty in Malawi over the last two decades as the state actively worked with the international agricultural research centres to promote hybrid maize as part of a strategy to increase food production (Harrigan 2008). In the 1980s, Malawian plant breeders, with support from USAID, the Rockefeller Foundation and the World Bank, developed and released a flint hybrid, which they named MH18. The NSCM, then a public corporation, promoted MH18 extensively as being more appropriate to smallholder farmer needs because of its yield, storage and food processing characteristics. In 1997, the E.U. provided credit for hybrid seed and fertilizer (Muhr et al. 1999), and in 1998 hybrid MH18 maize seed and fertilizer were distributed for free as a part of the Starter Pack. This distribution led to an increase in the area of hybrid maize planted throughout the country, from 7 percent in 1988 to over 30 percent in the next decade, and there were claims that a new "Green Revolution" had hit Malawi (Smale 1995).

This marriage between the state and private sector was short lived; in subsequent years, after acquiring the National Seed Company in 1996, Monsanto viewed MH18 as less profitable and took an active position *against* it.[6] As the major seed supplier in the country, Monsanto's decision to phase it out meant that MH18 was largely unavailable for several years either on the open market or as part of the subsidized Starter Pack. Instead, farmers were forced to rely on seeds, sold on the market by numerous foreign seed companies including Pioneer and Monsanto, that were bred in other countries and under different agroecological conditions.

Donors and the government also changed positions on hybrid versus OPV several times. For example, OPV was initially the only seed available in the Targeted Inputs Program, but later hybrid was also included and overshadowed OPV maize. For farmers who relied on the market or the Starter Pack for some of their seed, these contestations over maize variety have had profound implications for what they can plant.

The introduction of GM maize in research sites and potentially as a source of food aid in Malawi further erodes farmers' seed sovereignty. In Malawi and elsewhere in sub-Saharan Africa, Monsanto promotes Roundup Ready GM crops as a means to reduce labour requirements for HIV-infected households, despite the fact that HIV-affected rural households are often too poor to afford the herbicides and seeds (Bezner Kerr 2010). Monsanto also promotes seeds and herbicides in Malawi under the rubric of "conservation agriculture," in concert with the Ministry of Agriculture and NGOs such as Sasakawa (Bezner Kerr 2010). This technology, which involves using more herbicides to replace tillage, has in other countries led to greater uptake of herbicide-tolerant crops and, over time, an increase in herbicide-resistant weeds (Owen and Zelaya 2005; Sassenrath et al. 2008).

More recently Monsanto and other agricultural corporations have begun to invest in developing drought-tolerant genetic material. In one instance, Monsanto invested $750 million in collaboration with BASF in breeding drought-tolerant crops (Mitchell 2007), and in a collaboration with the Gates Foundation, the CGIAR international agricultural research system and the African Agricultural Technology Foundation, it is developing GM drought tolerant maize (Reuters 2008). Monsanto, BASF, Syngenta and other agri-businesses have also filed over five hundred patent documents for "climate-ready" genetic material (ETC Group 2008). All of these efforts point to increased corporate control over the types of seeds that farmers will plant in Malawi and elsewhere in the Global South. Ironically, they are embedded in competing discourse about avenues to address food security for the poor in the face of climate change (ETC Group 2008; Glover 2008).

Nonetheless, the behaviour of Monsanto and other agri-businesses in Malawi has not been as overtly controlling as in other places, where farmers must sign contracts that allow Monsanto to inspect their farm at any time (Kneen 1999). While in other countries Monsanto operates aggressively and litigiously to increase corporate control over seeds, in Malawi Monsanto works in concert with agro-input stores, NGOs and government agencies to distribute seeds, with no contractual relationships evident for commercialized GM crops as of 2009. Monsanto also donates money to non-profit organizations in Malawi, including World Vision, which along with other NGOs and government agencies received 700 metric tons of GM maize seed in 2005 from Monsanto (Monsanto Company 2005). This softer, philanthropic approach is in keeping with Monsanto's Smallholder Farmer Program, initiated in 2000 and aims to its their reach (and market share) with developing country farmers, increase support for biotechnology and improve its international image (Glover 2007, 2008).

A positive shift in seed access in Malawi is the formation of smallholder seed multiplication groups, which multiply and sell seed throughout the country, aiming to increase access to self-pollinating crops unavailable through the private sector. In practice thus far, the seed supply from these farmer groups has increased diversity of seed type in different regions of the country, but this supply is neither nationwide nor consistent and relies on donor support. There have also been efforts to link the formal breeding system with bean seed distribution by a consortium of the Ministry of Agriculture, NGOs and the Centre for Tropical International Agriculture, involving participatory plant breeding and distribution of selected farmer varieties to NGOs, hospitals and extension agents. While shown to be effective at increasing bean seed access in East Africa, it is not yet clear whether this approach will make a difference to peasant farmers in Malawi.

There are also pockets of hope for greater seed sovereignty where local seed systems are supported and celebrated. One NGO initiative, the Rumphi Food Security Project, in collaboration with the Ministry of Agriculture and local farmers, held a seed fair in 2009, which displayed numerous local varieties of maize, legumes and other crops. There are also numerous small community seed banks, such as the legume seed bank run by a nascent farmer association in Ekwendeni region, linked to efforts to promote legumes as an alternative to fertilizer (Bezner Kerr 2010). Although competing with much more organized and well-funded multinational seed organizations, these small-scale efforts provide some possible alternatives that may help lay the foundation for seed sovereignty for smallholder farmers in Malawi.

Conclusion

As we have seen, peasants in Malawi are losing their seed sovereignty through the incursion of the private sector into breeding and distribution of seed and the multiple links between the private sector, the state and aid organizations. Coupled with the push for a neoliberal model of the state, which has dramatically eroded state support for agriculture, these trends have led to limited or no availability of some seeds (e.g., groundnuts) and a push towards inappropriate varieties in the case of maize. At this point there are no clear strong national peasant organizations in Malawi attempting to increase seed sovereignty, although there are a number of smaller farmer organizations that hold possibilities. Community and kin networks remain a viable and important source of seed for many smallholder farmers, but these networks are fraught with contestations that leave landless peasants, young women and AIDS-affected families with less access and control over seed.

Seed sovereignty is one way that farmers can work to wrest back control over their food system. Seeds are the crucial resource in food production and a source of cultural exchange and significance, but they are also sources of genetic information and a commodity that has played a key part in strengthening the neoliberal food regime. Seed is the fundamental source of food for farmers reliant on own-farm production and as such is intimately linked to the concept of food sovereignty. The ways in which maize and groundnut seed are accessed, who controls seed, how it is selected and distributed and how struggles over seed have manifested in Northern Malawi highlight the importance of seeds as an essential element in the contemporary fight for food sovereignty by peasant movements worldwide.

Notes

1. The examples herein are drawn from hundreds of interviews, informal discussions and observations in this region over the past ten years, as well as archival research in the Malawi National Archives in Zomba. I work in collaboration with the Soils, Food and Healthy Communities project (www.soilandfood.org), a participatory project

aimed at improving peasant farmers' food security, nutrition and soil fertility in the Ekwendeni area of the Malawi, and their support is warmly acknowledged. Support from the International Development Research Centre and a Social Sciences and Humanities Research Council doctoral grant is gratefully acknowledged. I also thank Max Pfeffer, Rebecca Nelson, David Pelletier, Raj Patel, Hannah Wittman, Annette Desmarais and the anonymous reviewers for helpful feedback.

2.   In an interview in 2004, Malawi's chief maize breeder indicated that in order to fund breeding activities for drought resistant open-pollinated varieties (OPV) of maize, he relied heavily on the funds from private seed companies for seed certification, due to the lack of state funds.

3.   Farmers in fact can breed hybrid maize seed, and it is largely farmers who produce hybrid maize in Malawi, but in the current context, this breeding is carried out under the supervision of seed companies, agricultural scientists and technicians.

4.   Information about this period of the input subsidy program and the general agricultural scientific community at the time comes primarily from an informal interview with Dr. Sieglinde Snapp, March 18, 2005, as well as other published documents cited in text.

5.   SeedCo also produces OPV varieties, some of which were sold as part of the coupon system and some of which were being exported to Zimbabwe at the time of publication.

6.   Monsanto Malawi's market manager explicitly indicated in an interview that MH18 was not profitable enough, because it had been bread by the Ministry of Agriculture and thus Monsanto did not have patent rights on it; later the company cited Grey Leaf Spot as the official reason for phasing it out, but plant breeders indicated that it would be simple to breed an MH18 variety that is resistant to this disease (Bezner Kerr 2006).

References

Berry, V., and C. Petty. 1992. *The Nyasaland Survey Papers 1938–1943: Agriculture, Food and Health.* London: Academy Books Ltd.

Bezner Kerr, R. 2005a. "Food Security in Northern Malawi: Historical Context and the Significance of Gender, Kinship Relations and Entitlements." *Journal of Southern African Studies* 31, 1.

_____. 2005b. "Informal Labor and Social Relations in Northern Malawi: The Theoretical Challenges and Implications of Ganyu Labor for Food Security." *Rural Sociology* 70.

_____. 2006. "Contested Knowledge and Disputed Practice: Maize and Groundnut Seeds and Child Feeding in Northern Malawi." PhD dissertation, Cornell University, Development Sociology, Ithaca, NY.

_____. 2008. "Gender and Agrarian Inequality at the Local Scale." In S. Snapp and B. Pound (eds.), *Agricultural Systems: Agro-ecology and Rural Innovation.* San Diego: Elsevier Press.

_____. 2010. "The Land Is Changing: Contested Agricultural Narratives in Northern Malawi." In P. McMichael (ed.), *Contesting Development: Critical Struggles for Social Change.* New York: Routledge.

Bezner Kerr, R., L. Dakishoni, L. Shumba, R. Msachi, M. Chirwa. 2008. "'We Grandmothers Know Plenty': Breastfeeding, Complementary Feeding and the Multifaceted Role of Grandmothers in Malawi." *Social Science and Medicine* 66.

Bezner Kerr, R., and L. Shumba. 2007. "Preliminary Findings: AIDS-Affected Families and Agricultural Options." Unpublished report, Ekwendeni Hospital SFHC Project. Ekwendeni, Malawi.

_____. 2008. "Resilience and Struggle: Agricultural Issues for AIDS-Affected Farmers in Malawi." American Association for Geographers Annual Meeting. Boston, MA.

Bourdieu, P. 1977. Outline of a Theory in Practice. Cambridge: Cambridge University Press.

Brush, S.A. 2004. Farmers' Bounty: Locating Crop Diversity in the Contemporary World. New Haven: Yale University Press.

Cromwell, E., P. Kambewa, R. Mwanza, R. Chirwa and KWERA Development Centre. 2001. "Impact Assessment Using Participatory Approaches: 'Starter Pack' and Sustainable Agriculture in Malawi." Agricultural Research and Extension Network Paper 112.

Davison, J. 1997. Gender, Lineage and Ethnicity in Southern Africa. Boulder, CO: Westview Press.

Davy, W. 1912. "Annual Report on the DoA for the year ended March 31, 1912. Zomba: Government Printer.

Denning G., P. Kabambe, P. Sanchez, A. Malik, R. Flor, R. Harawa, P. Nkhoma, C. Zamba, C. Banda, C. Magombo, M. Keating, J. Wangila and J. Sachs. 2009. "Input Subsidies to Improve Smallholder Maize Productivity in Malawi: Toward an African Green Revolution." PLoS Biology 7.

Devereux S. 1997. "Household Food Security in Malawi." IDS Discussion Paper 362. Institute of Development Studies, Brighton.

_____. 2002. "Safety Nets in Malawi: The Process of Choice." Paper prepared for IDS conference Surviving the Present, Securing the Future: Social Policies for the Poor in Poor Countries. Institute of Development Studies, Brighton.

Dorward A., E. Chirwa, D. Boughton, E. Crawford, T.S. Jayne, R.J. Slater, V. Kelly and M. Tsoka. 2008a. "Towards 'Smart' Subsidies in Agriculture? Lessons from Recent Experience in Malawi." Natural Resource Perspectives 116.

Dorward A, E. Chirwa, V. Kelly, T.S. Jayne, R. J. Slater and D. Boughton. 2008b. Evaluation of the 2006/7 Agricultural Input Subsidy Programme, Malawi: Final Report. London: School of Oriental and African Studies.

Edriss, A., H. Tchale and P. Wobst. 2004. "The Impact of Labour Market Liberalization on Maize Productivity and Rural Poverty in Malawi." Available at <pasad.uni-bonn.de/Edriss_et_al_The_Impact.pdf>.

ETC Group. 2008. "Patenting the 'Climate Genes'... and Capturing the Climate Agenda." ETC Group Communique 99.

FAOSTAT (Food and Agricultural Organization Statistics Division). 2010. Available at <faostat.fao.org>.

Freeman, H.A., P.J.A. Van der Merwe, P. Subrahmayan, A.J. Chiyembekza and W. Kaguongo. 2002. "Assessing Adoption Potential of New Groundnut Varieties in Malawi." Experimental Agriculture 38.

Gladwin, C.H. 1992. "Gendered Impacts of Fertilizer Subsidy Removal Programs in Malawi and Vameroon." Agriculture Economics 7: 141–53.

Glover D. 2007. "Monsanto and Smallholder Farmers: A Case Study in CSR." Third World Quarterly 28.

_____. 2008. "Made by Monsanto: The Corporate Shaping of GM Crops as a Technology for the Poor." In STEPS Working Paper 11. Brighton: STEPS Centre.

Harrigan J. 2003. "U-Turns and Full Circles: Two Decades of Agriculture Reform in Malawi,

1981–2000." *World Development* 31, 5: 847–63.

_____. 2008. "Food Insecurity, Poverty and the Malawian Starter Pack: Fresh Start or False Start?" *Food Policy* 33.

Hotz, C., and R.S. Gibson. 2001. "Complementary Feeding Practices and Dietary Intakes from Complementary Foods amongst Weanlings in Rural Malawi." *European Journal of Clinical Nutrition* 55.

Kalinga, Owen J.M. 1993. "The Master Farmers' Scheme in Nyasaland 1950–1962: A Study of a Failed Attempt to Create a 'Yeoman' Class." *African Affairs* 92, 268: 367–87.

Kanyongolo, F.E. 2004. "Land Occupations in Malawi: Challenging the Neoliberal Legal Order." In S. Moyo and P. Yeros (eds.), *Reclaiming the Land: The Resurgence of Rural Movements in Africa, Asia, and Latin America*. London: Zed Books.

Kloppenburg, J.R. 2004. *First the Seed: The Political Economy of Plant Biotechnology*. Second edition. Madison: University of Wisconsin Press.

Kneen B. 1999. "Restructuring Food for Corporate Profit: The Corporate Genetics of Cargill and Monsanto." *Agriculture and Human Values* 16.

Kydd, J.G., and R.E. Christiansen. 1982. "Structural Change in Malawi since Independence: Consequences of a Development Strategy Based on Large-Scale Agriculture." *World Development* 10.

La Vía Campesina. 2007. "Seed Heritage of the People for the Good of Humanity: Women Seed Forum in South Korea." Seoul, South Korea.

Lele, U. 1989. "Structural Adjustment, Agricultural Development and the Poor: Lessons from the Malawian Experience." MADIA Discussion Paper 9. Washington, DC: World Bank.

Levy S., C. Barahona and B. Chinsinga. 2004. "Food Security, Social Protection, Growth and Poverty Reduction Synergies: The Starter Pack Programme in Malawi." *Natural Resource Perspectives* 95.

McCann, J.C. 2005. *Maize and Grace: Africa's Encounter with a New World Crop 1500–2000*. Cambridge: Harvard University Press.

McCracken, J. 1982. "Planters, Peasants and the Colonial State: The Impact of the Native Tobacco Board in the Central Province of Malawi." *Journal of Southern African Studies* 9, 2.

Mitchell P. 2007. "GM Giants Pair Up to Do Battle." *Nature Biotechnology* 25.

Monsanto Company. 2005. "Monsanto Assists 140,000 Farmers in Malawi With Hybrid Seed." Available at <prnewswire.com/cgibin/stories.pl?ACCT=104&STORY=/www/story/12-20-2005/0004237231&EDATE=>.

Muhr, L., S.A. Tarawali, M. Peters, and R. Schultze-Krafti. 1999. "Foragelegumes for Improved Fallows in Agropastoral Systems of Subhumid West Africa. I. Establishment, Herbage Yield and Nutritive Value of Legumes as Dry Season Forage." *Tropical Grasslands* 33, 4.

NSO and ORC Macro. 2001. "Malawi Demographic and Health Survey 2000." Zomba, Malawi and Calverton, U.S.: National Statistics Office and ORC Macro.

_____. 2005. *Malawi Demographic and Health Survey 2004*. Zomba, Malawi and Calverton, Maryland, U.S.: National Statistical Office of Malawi (NSO) and ORC Macro.

Owen, M.D.K., and I.A. Zelaya. 2005. "Herbicide-Resistant Crops and Weed Resistance to Herbicides." *Pest Management Science* 61.

Peters, P. 1996. "Failed Magic or Social Context? Market Liberalization and the Rural Poor in Malawi." Development Discussion Paper No. 562. Cambridge, MA: Harvard

Institute for International Development.

Pineyro-Nelson, A., J. Van Heerwaarden, et al. 2009. "Transgenes in Mexican Maize: Molecular Evidence and Methodological Considerations for GMO Detection in Landrace Populations." *Molecular Ecology* 18, 4.

Reuters. 2008. "The Chicago Council on Global Affairs Announces Grant from the Bill & Melinda Gates Foundation for Project on Global Agricultural Development." Available at <reuters.com/article/pressRelease/idUS151889+01-Oct-2008+PRN20081001>.

Rusike, J., and M. Smale. 1998. "Malawi." In M.L. Morris (ed.), *Maize Seed Industries in Developing Countries.* Boulder and Mexico D.F.: Lynne Rienner Publishers and CIM-MYT.

Sassenrath, G.F., P. Heilman, E. Luschei, G. L. Bennett, G. Fitzgerald, P. Klesius, W. Tracy, J.R. Williford and P.V. Zimba. 2008. "Technology, Complexity and Change in Agricultural Production Systems." *Renewable Agriculture and Food Systems* 23.

Smale, M. 1995. "'Maize is Life': Malawi's Delayed Green Revolution." *World Development* 23, 5.

Smale, M., and T.S. Jayne. 2003. "Maize in Eastern and Southern Africa: Seeds of Success in Retrospect." EPTD Discussion Paper No. 97, January.

Snapp, S., and M. Blackie. 2003. "Realigning Research and Extension to Focus on Farmers' Constraints and Opportunities." *Food Policy* 28.

Tindall, P. 1968. *History of Central Africa.* London: Longman.

Tripp R. 2001. *Seed Provision and Agricultural Development.* London: Overseas Development Institute.

Weis, T. 2007. *The Global Food Economy: The Battle for the Future of Farming.* London and New York: Zed Books.

11

Seed Sovereignty
The Promise of Open Source Biology

Jack Kloppenburg

> The seed has become the site and symbol of freedom in an age of manipulation and monopoly of its diversity. It plays the role of Gandhi's spinning wheel in this period of recolonization through free trade. The *charkha* (spinning wheel) became an important symbol of freedom because it was small; it could come alive as a sign of resistance and creativity in the smallest of huts and poorest of families. —Vandana Shiva (1997: 126)

From the wheat plains of Saskatchewan to the soy fields of Brazil's Mato Grosso, from the millet plots of Mali's Nyéléni to the rice paddies of the Philippines' Pampangan, the seed has become a prominent symbol of the struggle against the neoliberal project of restructuring the social and natural worlds around the narrow logic of the market. More than a symbol, however, the seed is also the very object and substance of that contest. As both a foodstuff and means of production, the seed sits at a critical nexus where contemporary battles over the technical, social and environmental conditions of production and consumption converge and are made manifest. Who controls the seed gains a substantial measure of control over the shape of the entire food system.

It therefore follows that if true food sovereignty is to be achieved, control over genetic resources must be wrested from the corporations and governments that seek to monopolize them and be restored to, and permanently vested in, social groups and/or institutions with the mandate to sustain them and to facilitate their equitable use. La Vía Campesina has recognized this necessity, identifying "seeds as the fourth resource ... after land, water and air" (La Vía Campesina 2001: 48) and declaring that "sustainability is completely impossible if the right of the peoples to recover, defend, reproduce, exchange, improve and grow their own seed is not recognized. Seeds must be the heritage of the peoples to the service of human kind" (La Vía Campesina 2009). That is, full realization of food sovereignty must be predicated on the attainment of what we may term "seed sovereignty."

How, then, can seed sovereignty be achieved in the current global, political-economic conjuncture? Those who believe that "another world is possible" face the two strategic tasks implied by Vandana Shiva in the quote above: the deployment of resistance against the project of neoliberalism and the creation of viable alternatives. On the one hand, a new set of global actors is beginning to resist the concentration of corporate power in the life sciences industry, the extension of intellectual property rights (IPRs), the privatization of public science, the spread of genetically modified crops, the development of "Terminator" technologies and the proliferation of bioprospecting/biopiracy. On the other hand, these global actors are also beginning to create spaces for the introduction and elaboration of alternatives such as farmers' rights, participatory plant breeding, a revitalized public science, the development of agroecology and support for decentralized and community-based seed distribution and marketing.

It is my contention that while resistance has often been effective, there is much more to be done in the realm of creating alternative spaces. This is especially important because the mechanisms that have been developed to address the inequities of such practices as bioprospecting have actually functioned to integrate farmers and indigenous peoples more closely into the existing market rather than to construct new and positive spaces for alternative action. Specifically, inasmuch as they have accepted the principle of exclusion — rather than sharing — as their constitutive basis, such arrangements have all proved inadequate even at defending, much less at reasserting or enlarging, peasant or community seed sovereignty. This chapter explores open source biology as a mechanism for simultaneously pursuing both effective resistance and the creation of a protected space into which practices and institutions with truly transformative capacity can be introduced and elaborated.

The Erosion of Farmers' Seed Sovereignty: The Privatization of Biodiversity

Until the 1930s, farmers worldwide enjoyed nearly complete sovereignty over their seeds, that is, they decided what seeds to plant, what seeds to save and who else might receive or be allocated their seed as either food or planting material. Such decisions were made within the overarching norms established by the cultures and communities of which they were members. While these customary arrangements often recognized some degrees of exclusivity in access to genetic resources, they were largely open systems that operated on the bases of reciprocity and gift exchange rather than the market. Indeed, these customary arrangements usually functioned to stimulate and facilitate — rather than restrict — the wide dissemination of seed (Zimmerer 1996; Brush 2004; Salazar et al. 2007). The sharing of seed resulted in the continuous recombination of genetic material, which in turn produced the agronomic resilience that is characteristic of peasant- and farmer-developed crop varieties and landraces. This historic creation and recreation of crop diversity not only fed particular communities and peoples but also collectively constitutes the genetic foundation on which future world food production can most sustainably and equitably be based.

Since the 1930s, farmers' sovereignty over seeds has been continuously and progressively eroded while the sovereignty of what is now a "life sciences industry" has been correspondingly enlarged. The development of inbreeding/hybridization in the 1930s first separated the farmer from the effective reproduction of planting material and created the opening needed for private capital to profit in the seed sector. Seed companies then used their increasing influence to obtain "plant breeders' rights": legislation that conferred exclusive control to them over varieties in crops in which hybridization was not possible (Kloppenburg 2004).

Subsequently, the seed industry has pursued both of these routes — technical and social — to further restrict farmers' access to seed to the confines of an increasingly narrow set of market mechanisms. The structures of science have been used to develop "Terminator" and "Transcontainer" technologies, which genetically sterilize seed in order to prevent plant-back by farmers. Both national and international structures of governance — that is, institutions such as the World Trade Organization and the Convention on Biodiversity, as well as national legislatures — have been used for the global elaboration of a set of intellectual property rights based on the principle of *exclusion*. By making saving of patented seed illegal, these arrangements are effectively an enclosure of farmers' practices as well as their seed.

These technical and social processes of commodification are enabled in important ways by two key features of the organization of knowledge production and accumulation in the plant sciences. First, the development of agronomically useful and novel (and therefore patentable) plant varieties has been predicated on access by breeders to the enormous pool of biodiversity that has been produced and reproduced by farmers and indigenous peoples. Systematic appropriation of landraces from farming communities by university and government scientists, their storage in genebanks controlled by governments, corporations and non-governmental organizations and their subsequent use in breeding programs is a long standing practice. This bioprospecting has increasingly been understood as "biopiracy" insofar as no or insufficient benefits flow reciprocally to the communities and peoples who freely shared the collected materials as the "common heritage of humankind" (Mgbeoji 2006; Mushita and Thompson 2007).

Second, the supplanting of classical crop breeding by transgenic methods, the progressive weakening of public research institutions such as universities, government and international facilities and the subordination of their work to corporate objectives has resulted in an overwhelming focus on the development of genetically modified varieti s (Knight 2003; Gepts 2004). After twenty years and billions of dollars of expenditures, GMO cultivars incorporate only two traits (one being herbicide resistance) in only four crops (maize, soy, cotton and canola). The subsequent failure of public science to provide an alternative to this narrow range of patented, corporate seeds has permitted the global dissemination of crop varieties that do not meet the needs of most farmers, that often cannot be legally saved, that reinforce the expansion of unsustainable monocultures and that contaminate other varieties with proprietary transgenes (Quist and Chapela 2001; Rosset 2006).

Seed sovereignty has been gradually transferred from farmers and their communities to the boardrooms of the five transnational firms known as the "Gene Giants" for their domination of the U.S. $20 billion annual global market for seeds. Once freely exchanged according to an ethic of sharing, access to seeds is now ruled by a set of legal mandates based on the principle of exclusion. Once bred by farmers to meet local needs, seeds are now genetically engineered by corporate scientists to the specifications of a globally distributed industrial agriculture geared not to feeding people but to feeding the corporate bottom line.

Resisting Exclusion, Creating Alternatives?

An encouraging feature of the past decade has been the emergence of a robust, globally distributed resistance to the ways in which capital has chosen to shape global agricultural markets, develop biotechnology and construct IPRs (Schurman and Kelso 2003). Widespread popular aversion to patents on life forms and to such pernicious applications as Terminator technology has been joined to concerns in the scientific community about growing limits on their own freedom to operate amongst the proliferating corporate patent thickets. Peasants, farmers, indigenous peoples and civil-society advocacy groups have been working as part of a diffuse but powerful social movement that has had success at slowing — though certainly not stopping — what has come to be broadly understood as the project of corporate globalization in agriculture. With the emerging crises in environment, energy and food production, we can anticipate growing resistance and the opening of space for the pursuit of "another world."

Resistance activities have shown increasing numbers of people that another world is necessary; it is also critical to show them that another world is actually possible. It is this creative arena that needs to be strengthened. To date, three principal approaches to the protection of genetic resources have been pursued by farmers, indigenous peoples and advocacy groups. These are the establishment of farmers' rights at the international level, attempts to embed traditional resource rights in national-level IPR legislation and the promulgation of a wide range of bilateral agreements between bioprospectors and target communities themselves.

Much of the affirmative action that has been pursued on genetic resources over the last twenty-five years has been undertaken under the rubric of the construct called farmers' rights. Written into the 1989 agreed interpretation of the FAO International Undertaking on Plant Genetic Resources, farmers' rights were to have balanced patent-like plant breeders' rights by conferring on farmers and indigenous peoples a moral and material recognition of the utility and value of the labour they have expended, and continue to expend, in the development and regeneration of crop genetic diversity. Alas, farmers' rights as they have appeared in international fora have been little more than a rhetorical sleight of hand, a means of diverting activist energies into prolonged discussions with the corporate/bureaucratic masters of passive-aggressive negotiation. For example, the final result of twelve years of talks in the FAO was, in 2001, approval of an International Treaty on Plant Genetic Resources for Food and Agriculture (FAO 2001). The treaty acknowledges the rights of farmers to "save, use, exchange, and sell farm saved seed," but renders this a privilege "subject to national legislation," which is to say those rights are subordinated to — and thus negated by — conventional IPR rules such as patenting.

A second line of action has involved efforts to exploit an opening in the WTO's Agreement on Trade-Related Aspects of Intellectual Property Rights (TRIPS). Article 27.3(b) of TRIPS requires WTO member nations to offer some form of intellectual property rights in plants through patenting, plant breeders' rights or an effective *sui generis* system. In theory this option provides nation states with an opportunity to shape legislation to protect the interests and needs of farmers and indigenous peoples. In practice, many nations — often under pressure from the U.S. and other advanced capitalist nations — simply adopt a conventional plant breeders' rights framework that provides patent-like protection for plant breeders but fails to provide symmetrical rights for farmer-developed cultivars. Genetic Resources Action International (GRAIN 2003) has documented over twenty-five instances of such legislative action in countries of the Global South.

With international and national-level institutions insufficiently attentive to their needs and rights, communities of peasants, farmers and indigenous peoples have in many cases turned to a third mechanism — direct bilateral arrangements — in an effort to establish rights over crop biodiversity, manage bioprospecting and derive a flow of benefit from genetic materials. These have ranged from detailed and highly legalistic models typical of western patent law to frameworks that are more like a treaty than a contract (Posey and Dutfield 1996; Marin 2002). The evidence produced by a number of assessments of these arrangements shows that not only have they failed to deliver any significant benefits, they have also frequently caused considerable social disruption and too often actually actively damaged the contracting communities (Brown 2003).

Hayden (2003: 233) describes the "noisy demise" of the "debacle" of bioprospecting among the Maya of Chiapas, and Greene (2004: 104) documents the "rather extraordinary mess" that resulted from the dissolution of a similar project among the Aguaruna of Peru. Both projects floundered and ultimately failed due to the inability of the ethnobiologists doing the collecting to establish consent and compensation arrangements that were broadly acceptable to the indigenous communities involved. Among the Aguaruna, the pivotal issue had to do with the inadequacy of the contracted royalty rates, while among the Maya the central issue was the management and control of the NGO that was created to distribute possible royalties. In both cases, the multiple ethical, representational and financial dilemmas raised by these and other bioprospecting projects remain unresolved and perhaps unresolvable.

It should not be surprising that these three modalities discussed above have failed at ensuring peasants, farmers and indigenous peoples' equitable rights to genetic resources. The existing IPR regime is a juridical construct shaped to serve corporate interests. Moreover, the collective character of the production of crop genetic resources and their wide distribution and exchange almost always makes appropriate allocation of "invention" to a person, persons, a community, communities or even a people or peoples an impracticable — and often divisive — task (Kloppenburg and Balick 1995; Brush 2004). Even if some legitimate partner can be identified, it is difficult to see how peasants and indigenous peoples can provide informed consent to bioprospecting activities and construct exchange agreements adequately sensitive to their own interests. Further, the indeterminacy of the value of any material at the point of collection, the difficulty of distinguishing the magnitude of value added in subsequent breeding and marketing and the imbalance of power between donor and collector render the flow of any material benefit via instruments such as access fees, licensing fees and royalties uncertain at best.

Beyond these practical difficulties, there is a larger issue. The nature of property is called into question when some farmers and indigenous peoples reject the very notion of owning seeds or plants, which they may regard as sacred or as a collective heritage (Hurtado 1999; Salazar et al. 2007). IPRs are actually a means of circumventing and obscuring the reality of *social* production and subsuming the products of social production under private ownership for the purposes of *excluding* others from use. How can they be anything but antagonistic towards social relations founded on cooperative, collective, multigenerational forms of knowledge production?

If food sovereignty is going to be possible, might its development not be facilitated more by the expansion of opportunities for humans to enact the principle of sharing than by the extension of the reach of the principle of exclusion? An alternative route to establishment of a just regime for managing flows of crop germplasm might involve creation of a mechanism for germplasm exchange that allows sharing among those who will reciprocally share, but excludes those who will not: a protected commons.

I suggest that "open source biology" offers the means to establish and elaborate such a protected commons for crop genetic resources. While it is no panacea, it represents a plausible mechanism for engaging in both resistance and creativity and for moving in concrete ways towards realization of seed sovereignty.

Open Source Movements: From Software to Wetware

Though to them it may seem so, farmers, peasants and indigenous peoples are not the only targets of what McMichael (1996: 31) calls "the globalization project" and what Hardt and Negri (2000) name as "Empire." But peasants, farmers and indigenous peoples may find resources for their own struggles in the parallel experiences of others. And so it is with seeds and software.

Nowhere have the issues of commodification, ownership and exclusion of use been played out more clearly than in the field of software development. Advances in hard and soft digital technologies have galvanized the rapid emergence of productive sectors of enormous power and value. Although creative capacity in software development is globally distributed among individuals, universities and variously sized firms, a few companies have attained a dominant market position from which they have used copyright and patent arrangements to reinforce their own hegemony by restricting the use of their proprietary software, especially of operating system code. Frustrated by these expanding constraints on their ability to add to, and to modify and to share as freely as seemed personally and socially desirable, software developers have sought ways to create space in which they can develop content and code that can be liberally exchanged and built upon by others.

The resultant emergence of a dynamic "free and open source software" (FOSS) movement has been widely documented and analyzed (Raymond 1999; Stallman 2002; Weber 2004). The FOSS movement is quite diverse, encompassing a considerable range of organizations and methods (e.g., Creative Commons, FOSSBazaar, Free Software Foundation, Open Source Initiative). What unifies these initiatives is a commitment to allowing software users to access and modify code and, critically, to implementing an enforceable legal framework that preserves access to the original source code and to any subsequent modifications and derivatives.

Software released under open source arrangements is copyrighted and made freely available through a licence that permits modification and distribution as long as the modified software is distributed under the same licence through which the source code was originally obtained. That is, source code and any modifications must be freely accessible to others (hence "open source") as long as they in turn agree to the provisions of the open source licence. Note that the "viral" effect of such "copyleft" arrangements enforces continued sharing as the program is disseminated. Just as importantly, this form of licensing also prevents appropriation by companies that would make modifications for proprietary purposes since any software building on the licensed code is required to be openly accessible. Thus, software developed under open source arrangements is released not into an open access commons but into a protected commons populated by those who agree to share.

The FOSS movement has enjoyed considerable success. Thousands of open source programs are now available, the best known among them being the operating system Linux. The originator of this program is Linus Torvalds, whose express objective was to develop a functional computer operating system as an alternative to those offered by Microsoft and Apple. Realizing that he could not undertake so large a task on his own, he released the "kernel" code of the program under an open source licence and asked the global community of programmers to contribute their time and expertise to its elaboration, improvement and modification. He subsequently involved thousands of colleagues in an ongoing, interactive process that has made Linux and its many iterations and flavours an operating system that competes with those of Microsoft and Apple.

The practical utility of this collective enterprise is captured in what is known as "Linus' Law": "Given enough eyeballs, all bugs are shallow" (Raymond 1999: 30). That is, the mobilization of large numbers of people working freely together in "decentralized/distributed peer review" generates what Eric Raymond (1999: 31) calls a "bazaar" — as opposed to a "cathedral-builder" — approach to innovation. Users are transformed from customers into co-developers, and the capacity for creative, rapid, site-specific problem solving is greatly multiplied. The social utility of such a collective enterprise is that, subsequent to the open source licensing arrangements under which work proceeds, the results of social labour remain largely socialized and cannot be monopolized.

That they cannot be monopolized does not mean that they cannot be commercialized. Many of the programmers working on open source projects are motivated by peer recognition and the opportunity to contribute to the community (Raymond 1999: 53). But labour can (and should) also be materially rewarded. As the Free Software Foundation (2008) has famously observed, "Free software is a matter of liberty, not price. To understand the concept, you should think of free as in free speech, not as in free beer." Open source software need not be made available at no cost, but it must be available free of restrictions on further expression via derivative works.

A number of analysts have begun to look to the FOSS movement as a model for development of "open source biology" practices — "BioLinuxes" (Srinivas 2006) — that might be the basis for resisting enclosure of the genescape and for reasserting modalities for freer exchange of biological materials and information (Deibel 2006; Rai and Boyle 2007; Hope 2008). Efforts have been made to apply open source and copyleft principles to a variety of bioscience enterprises, including mapping of the haplotypes of the human genome (International HapMap Project), drug development for neglected diseases in the Global South (the Tropical Diseases Initiative), the standardization of the components of synthetic biology (BioBricks Foundation) and a database for grass genomics (Gramene).

By far the most substantial of such initiatives has been that undertaken by Richard Jefferson and his colleagues at the non-profit CAMBIA. Convinced of the utility of advanced genetics for improving agriculture in marginal and inadequately served communities, Jefferson had been frustrated by the narrow uses to which corporations have put genetic engineering and deeply critical of the constraints they place on the sharing of patented technology (Poynder 2006). With the explicit intent to extend the metaphor and concepts of open source to biotechnology, Jefferson has fostered the construction of Biological Open Source (BiOS), an "innovation ecosystem" designed to "ensure common access to the tools of innovation, to promote the development and improvement of those tools, and to make such developments and improvements freely accessible to both academic and commercial parties" (BiOS 2009a). BiOS involves integrating cutting-edge biological research with open source licensing arrangements that "support both freedom to operate, and freedom to cooperate" in a "protected commons" (BiOS 2009b).

A BioLinux for Seeds?

The seed sector appears to offer some interesting potentials for elaboration of a "BioLinux" approach to open source innovation (Douthwaite 2002; Srinivas 2006; Aoki 2008). Millions of peasants, farmers and indigenous communities the world over are engaged in the recombination of plant genetic material and are constantly selecting for improvements. Even more massively than their software hacker counterparts, they are effectively participating in the process of distributed peer production that Eric Raymond has characterized as the "bazaar." Like programmers, farmers have found their traditions of creativity and free exchange being challenged by the IPRs of the hegemonic "permission culture" and have begun looking for ways not just to protect themselves from piracy or enclosure but also to reassert their own norms of reciprocity and innovation.

Moreover, peasants, farmers and indigenous communities have potential allies in this endeavour who themselves are capable of bringing useful knowledge and significant material resources to bear. Although its capacity is eroding, public plant breeding still offers an institutional platform for developing the technical kernels needed to galvanize recruitment to a protected commons. And in the practice of participatory plant breeding there is an extant vehicle for articulating the complementary capacities of farmers and scientists in the North as well as the South (Almekinders and Jongerden 2002; Murphy et al. 2004; Salazar et al. 2007). Could copyleft arrangements establish a space within which these elements might coalesce and unfold into something resembling seed sovereignty?

The recent appreciation of the potential utility of open source methods for the seed sector was preceded by a similar apprehension on the part of a member of the plant breeding community itself. At the 1999 Bean Improvement Conference, University of Guelph bean breeder Tom Michaels presented a paper titled "General Public License for Plant Germplasm." In it, he noted that as a result of

> the opportunity to obtain more exclusive novel gene sequence and germplasm ownership and protection, the mindset of the public sector plant breeding community has become increasingly proprietary. This proprietary atmosphere is hostile to cooperation and free exchange of germplasm, and may hinder public sector crop improvement efforts in future by limiting information and germplasm flow. A new type of germplasm exchange mechanism is needed to promote the continued free exchange of ideas and germplasm. Such a mechanism would allow the public sector to continue its work to enhance the base genotype of economically important plant species without fear that these improvements, done in the spirit of the public good, will be appropriated as

part of another's proprietary germplasm and excluded from unrestricted use in other breeding programs. (1999: 1)

The specific mechanism Michaels goes on to propose is a general public license for plant germplasm (GPLPG) that is explicitly modelled on a type of licence common to open source arrangements in software. This mechanism is simple, elegant and effective. It can be used by many different actors (individual peasants, farmers, communities, indigenous peoples, plant scientists, universities, NGOs, government agencies and private companies) in many places and diverse circumstances. Properly deployed, it could be an effective mechanism for creating a protected commons for those who are willing to freely share continuous access to a pool of plant germplasm for the purposes of bazaar-style, distributed peer production.

Implementation of open source mechanisms such as the GPLPG could have significant effects consistent with both strategies of resistance and creativity. In terms of resistance, the GPLPG would:

Prevent or impede the patenting of plant genetic material. A GPLPG would not directly prohibit patenting (or any other form of IPR protection) of plant genetic material but would render such protection pointless. The GPLPG mandates sharing and free use of the subsequent generations and derivatives of the designated germplasm. In effect, this prevents patenting since there can be no income flow from the restricted access to subsequent generations and derivative lines that it is the function of a patent to generate. Further, the viral nature of the GPLPG means that as germplasm is made available under its provisions and used in recombination, there is a steadily enlarging the pool of material that is effectively insulated from patenting. Enforcing the GPLPG against possible violators would not be easy given the resources necessary. But even the mere revelation of violations would have the salutary effect of illuminating corporate malfeasance and eroding the legitimacy of industry and its practices.

Prevent or impede bioprospecting/biopiracy. The GPLPG could be similarly effective in deterring biopiracy. Faced with a request to collect germplasm, any individual, community or people could simply require use of a materials transfer agreement incorporating the GPLPG provisions. Few commercially oriented bioprospectors will be willing to collect under those open source conditions. Again, enforcing the GPLPG against possible violators would not be easy, but instances in which bioprospecting can be revealed to unambiguously be biopiracy would contribute to public awareness and strengthen popular and policy opposition to unethical appropriation of genetic resources.

Prevent or impede the use of peasant- and farmer-derived genetic resources in proprietary breeding programs. Because neither the germplasm received under a GPLPG nor any lines subsequently derived from it can be use-restricted, such materials are of little utility to breeding programs oriented to developing proprietary cultivars. Any mixing of GPLPG germplasm with these IPR-protected lines potentially compromises their proprietary integrity. Application of the GPLPG to landraces could therefore effectively prevent their use in proprietary breeding programs.

Prevent or impede further development and deployment of GMOs. The development of transgenic cultivars almost universally involves multiple layers of patented and patent-licensed germplasm. Moreover, all of the critical enabling technologies employed in genetic engineering are patented and their use restricted by licences. Given the large investments that have been made and accompanying expectations of high financial returns, GMOs will not be developed if they cannot be IPR-protected. Any mixing of GPLPG germplasm with these IPR-protected materials and tools potentially compromises their proprietary status. Use of the GPLPG cannot itself stop the further development of GMOs, but it can impede it by preventing additional genetic resources from being drawn into the web of proprietary and IPR-protected materials.

In addition to its capacity for reinforcing resistance, the GPLPG may have even more potential for creativity, for the creation of effective space for the elaboration of transformative alternatives. For example, implementation of the GPLPG would help to:

Develop a legal/institutional framework that recognizes peasants', farmers' and indigenous peoples' collective sovereignty over seeds. A major advantage of the GPLPG is that it does not require the extensive development of new legal statutes and institutions for its implementation. It relies on the simple vehicle of the materials transfer agreement, which is already established and enforceable in conventional practice and existing law. It uses the extant property rights regime to establish rights over germplasm, but then uses those rights to assign sovereignty over seed to an open-ended collectivity whose membership is defined by the commitment to share the germplasm they now have and the germplasm they will develop. Those who do not agree to share are self-selected for exclusion from that protected commons.

Develop a legal/institutional framework that allows peasants, farmers and indigenous peoples to freely exchange, save, improve and sell seeds. For farmers, the feature of the space created by implementation of the GPLPG that is of principal importance is the freedom to plant, save, replant, adapt, improve, exchange, distribute and sell seeds. The flip side of these freedoms is responsibility (and under the GPLPG, the obligation) to grant others within the collectivity the same freedoms; no one is entitled to impose purposes on others or to restrict the range of uses to which seed might be put. In the face of increasing restrictions on their degrees of freedom to access and use seed, application of the GPLPG offers a means for farmers to create a semi-autonomous, legally secured, protected commons in which they can once again work collectively to express the inventiveness that has historically so enriched the agronomic gene pool.

Develop an institutional framework in which peasants, farmers and indigenous peoples cooperate with plant scientists in the development of new plant varieties that contribute to a sustainable food system. The protected commons that could be engendered by the GPLPG can, and must, also encompass scientific plant breeders whose skills are different from, but complementary to, those of farmers. Many new cultivars will be needed to meet the challenges of sustainably and justly feeding an expanding global population in a time of energy competition and environmental instability. The open source arrangements that have undergirded the successes of distributed peer production in software could have a similar effect in plant improvement. If in software it is true that "to enough eyes, all bugs are shallow," it may follow that "to enough eyes, all agronomic traits are shallow." Participatory plant breeding offers a modality through which the labour power of millions of farmers can be synergistically combined with the skills of a much smaller set of plant breeders. The GPLPG offers plant scientists in public institutions a means of recovering the freedoms that they — no less than farmers — have lost to corporate penetration of their workplaces. Public universities, government agencies and the Consultative Group on International Agricultural Research (CGIAR) system should be the institutional platform for knowledge generation based on the principle of sharing rather than exclusion. Public plant breeders, too, can be beneficiaries of and advocates for the protected commons.

Develop a framework for marketing of seed that is not patented or use-restricted. The GPLPG is antagonistic not to the market but to the use of IPRs to extract excess profits and to constrain creativity through restrictions on derivative uses. Under the GPLPG, seed may be reproduced for sale and sold on commercial markets. By carving out a space from which companies focusing on proprietary lines are effectively excluded, the GPLPG creates a market niche that can be filled by a decentralized network of small-scale, peasant/farmer-owned and cooperative seed companies that do not require large margins and that serve the interests of seed users rather than investors.

Seed sovereignty need not involve peasants, farmers and indigenous peoples alone, nor can it be achieved solely by these social actors. Seed sovereignty will be manifested as a system encompassing peasants, farmers, indigenous peoples, plant scientists, public scientific institutions and seed marketers. GPLPG/BioLinux/open source/copyleft arrangements could plausibly constitute a legal/regulatory framework that could open an enabling space within which all these different social actors could be effectively affiliated.

Pursuing Seed Sovereignty

> We should sit down with the legal people who drew up the Creative Commons
> licenses and see whether farmers could use a similar approach with seeds.
> —José Bové (2005: 11)

If seed sovereignty is to be pursued as part of a larger conception of food sovereignty, what is to be done? José Bové is clear about what path should be taken. If germplasm had been made available by peasants, farmers and indigenous peoples under the public licensing of the kind described above since 1950, world agriculture would look very different today. At a minimum, the public agricultural research system would be far more robust than it is now, most seeds in most genebanks would be freely available to any breeders willing to share the results of their work and it would be Monsanto — not peasants farmers and indigenous peoples — that would be finding the international plant genetic resources regime to be unduly restrictive. With such potency, might a BioLinux approach be useful today?

A wide variety of analysts have grappled with what to do about the asymmetric and unjust character of plant germplasm use and exchange. Their counsel can be separated into three types. The first is to do nothing. Some are so overwhelmed by practical complexities and moral ambiguities that they simply don't know what to do and fail to provide any effective guidance at all (e.g., Brown 2003; Gepts 2004; Eyzaguirre and Dennis 2007). Others bemoan the problematics of existing arrangements but accept their inevitability (e.g., Wright 1998; Fowler 2003; Brush 2007). Dusting off an old seed industry apologia, for example, Brush (2007: 1511) concludes that existing mechanisms of development assistance and technology transfer represent sufficient means of ensuring reciprocity and benefit sharing. Fowler (2003: 3, 11) flatly declares that "for better or worse, the debate concerning whether the international community will sanction the existence and use of IPRs in relation to germplasm… is over" and that "anyone who is not happy will remain unhappy."

The second and much larger group agrees that something needs to be done about the injustices but that the realities of corporate power and a dominant capitalism require a "situational pragmatism" (Brown 1998: 205) that involves cutting the best deal possible. So Mgbeoji (2006: 170) recommends that indigenous peoples consider a "more astute and pragmatic response" to patenting of sacred plants. Salazar et al. (2007) advise trying out the new and trendy "declaration of origin" as a means of preventing appropriation. This is the well worn terrain of all the bioprospecting contracts and the discoverer's rights and the geographic indications and the biopartnerships and the recognition funds and the royalty agreements and the exploration fees and the all the other arrangements that have been proposed and tried.

I have no objection to trying them out and am in no position to tell any peasant communities or indigenous peoples what they should or should not do. I will point out, however, that none of these arrangements have yet worked, largely because of the erosive effects that inevitably accompany a compensationist, exclusionist articulation to the market. Darryl Posey observed that, as far as he was concerned, these deals were holding actions that would not enfranchise anyone but that would "at least buy some time" (cited in Hayden 2003: 38). But, buy time for what? Hurtado (1999: 7–8) warns of the dangers in the pressures to be pragmatic and to accept what he calls the "intermediate" solutions where

> we must not go to extremes, but rather negotiate and arrive at a mid-point. And in this the INTERMEDIATES are the special or *sui generis* regimes, which seek to sit indigenous people at the negotiating tables, in order to talk us into submission. Because it is there where the banana skins are placed, it is there where we start to skid.

A BioLinux or other sharing arrangement may be a viable third option. The aggressions of the neoliberal project must, of course, be resisted whenever possible. However, resistance is a necessary but not sufficient condition for the realization of seed sovereignty (or, for that matter, of food sovereignty). Resistance, complemented by creative actions that are not just reactions to corporate/neoliberal conditions but which are offensive, affirmative, positive, pro-active undertakings designed to establish and maintain alternative, (relatively) autonomous spaces, has more potential for transformation.

Achieving seed sovereignty will not be easy. What is required is simultaneous and linked development of concepts and applications among peasants, farmers, indigenous communities, plant scientists, seed vendors, public institutions and civil-society advocacy groups in the face of corporate and state opposition. While open source biology is no cure-all, it may be a plausible vehicle for enacting the elements of resistance and creativity that La Vía Campesina and others suggest will be necessary for the achievement of seed sovereignty. Should we not, therefore, take the advice offered by José Bové? If there is to be food sovereignty, surely it will be facilitated and enabled by a struggle for seed sovereignty.

Note

A revised version of this chapter appears as "Impending Dispossession, Enabling Repossession: Biological Open Source and the Recovery of Seed Sovereignty" in *Journal of Agrarian Change* 10, 3 (2010).

References

Almekinders, C., and J. Jongerden. 2002. "On Visions and New Approaches: Case Studies of Organisational Forms in Organic Plant Breeding and Seed Production." Working Paper, Technology and Agrarian Development, Wageningen University, Netherlands.

Aoki, K. 2008. *Seed Wars: Controversies and Cases on Plant Genetic Resources and Intellectual Property*. Durham, NC: Carolina Academic Press.

BiOS (Biological Open Source). 2009a. "BiOS Material Transfer Agreement For Seeds/ Propagules." Available at <bios.net/daisy/bios/3004.html>.

_____. 2009b. "About BiOS (Biological Open Source) Licenses and MTAs." Available at <bios.net/daisy/bios/mta.html>.

Bové, J. 2005. "Convergence Zone: José Bové." *Seedling* (October).

Brown, M.F. 2003. *Who Owns Native Culture?* Cambridge, MA: Harvard University Press.

_____. 1998. "Can Culture Be Copyrighted?" *Current Anthropology* 39.

Brush, S.B. 2004. *Farmers' Bounty: Locating Crop Diversity in the Contemporary World*. New Haven, CT: Yale University Press.

_____. 2007. "Farmers' Rights and Protection of Traditional Agricultural Knowledge." *World Development* 35, 9.

Deibel, E. 2006. "Common Genomes: Open Source in Biotechnology and the Return of the Commons." *Tailoring Biotechnologies* 2, 2.

Douthwaite, B. 2002. *Enabling Innovation: A Practical Guide to Understanding and Fostering Technical Change*. Boston, MA: Zed Books.

Eyzaguirre, P.B., and E.M. Dennis. 2007. "The Impacts of Collective Action and Property Rights on Plant Genetic Resources." *World Development* 35, 9.

FAO (Food and Agriculture Organization of the United Nations). 2001. "International Treaty on Plant Genetic Resources for Food and Agriculture." Available at <fao.org/ ag/cgrfa/itpgr.htm>.

Fowler, C. 2003. "The Status of Public and Proprietary Germplasm and Information: An Assessment of Recent Developments at FAO." *IP Strategy Today* No. 7-2003. Ithaca, NY: *bio*Developments-International Institute.

Free Software Foundation. 2008. "Free Software Definition." Available at <gnu.org/philosophy/free-sw.html>.

Gepts, P. 2004. "Who Owns Biodiversity, and How Should the Owners Be Compensated." *Plant Physiology* 134 (April).

GRAIN. 2003. "Farmers' Privilege under Attack." Available at <grain.org/bio-ipr/?id=105>.

Greene, S. 2004. "Indigenous People Incorporated? Culture as Politics, Culture as Property in Biopharmaceutical Bioprospecting." *Current Anthropology* 45, 2 (April).

Hardt, M., and A. Negri. 2000. *Empire*. Cambridge, MA: Harvard University Press.

Hayden, C. 2003. *When Nature Goes Public: The Making and Unmaking of Bioprospecting in Mexico*. Princeton, NJ: Princeton University Press.

Hope, J. 2008. *Biobazaar: The Open Source Revolution and Biotechnology*. Cambridge, MA: Harvard University Press.

Hurtado, L.M. 1999. *Access to the Resources of Biodiversity and Indigenous Peoples*. Occasional Paper of the Edmonds Institute, Edmonds, WA. Available at <http://www.edmonds-institute.org/muelaseng.html>.

Kloppenburg, J. Jr. 2004. *First The Seed: The Political Economy of Plant Biotechnology, 1492–2000*. Madison, WI: University of Wisconsin Press.

Kloppenburg, J. Jr., and M. Balick. 1995. "Property Rights and Genetic Resources: A Framework for Analysis." In Michael Balick and Sarah Laird (eds.), *Medicinal Resources of the Tropical Forest: Biodiversity and Its Importance to Human Health*. New York: Columbia University Press.

Knight, J. 2003. "A Dying Breed: Public-Sector Research into Classical Crop Breeding Is

Withering." *Nature* 421 (February 6).

La Vía Campesina. 2001. "The Position of Vía Campesina on Biodiversity, Biosafety and Genetic Resources." *Development* 44, 4.

_____. 2009. "Food Sovereignty: A New Model for a Human Right." Available at <viacampesina.org/main_en/index.php?option=com_content&task=view&id=730&Itemid=3.

Marin, P.L.C. 2002. *Providing Protection for Plant Genetic Resources: Patents, Sui Generis Systems, and Biopartnerships.* The Hague: Kluwer Law International.

McMichael, P. 1996. "Globalization: Myths and Realities." *Rural Sociology* 61, 1.

Mgbeoji, I. 2006. *Global Biopiracy: Patents, Plants, and Indigenous Knowledge.* Ithaca, NY: Cornell University Press.

Michaels, T. 1999. "General Public License for Plant Germplasm: A Proposal by Tom Michaels." Paper presented at the 1999 Bean Improvement Cooperative Conference, Calgary, Alberta.

Murphy, K., D. Lammer, S. Lyon, B. Carter and S.S. Jones. 2004. "Breeding for Organic and Low-input Farming Systems: An Evolutionary-Participatory Breeding Method for Inbred Cereal Grains." *Renewable Agriculture and Food Systems* 20, 1.

Mushita, A., and C.B. Thompson. 2007. *Biopiracy of Biodiversity: Global Exchange as Enclosure.* Trenton, NJ: Africa World Press.

Posey, D.A., and G. Dutfield. 1996. *Beyond Intellectual Property: Toward Traditional Resource Rights for Indigenous Peoples and Local Communities.* Ottawa: IDRC.

Poynder, R. 2006. "Interview with Richard Jefferson." Available at <poynder.blogspot.com/2006/09/interview-with-richard-jefferson.html>.

Quist, D., and I.H. Chapela. 2001. "Transgenic DNA Introgressed Into Traditional Maize Landraces in Oaxaca, Mexico." *Nature* 414 (November 29).

Rai, A., and J. Boyle. 2007. "Synthetic Biology: Caught between Property Rights, the Public Domain, and the Commons." *PloS Biology* 5: 3 (March).

Raymond, E.S. 1999. *The Cathedral & the Bazaar: Musings on Linux and Open Source by an Accidental Revolutionary.* Sebastopol, CA: O'Reilly Media.

Rosset, P. 2006. "Genetically Modified Crops for a Hungry World: How Useful Are They Really?" *Tailoring Biotechnologies* 2: 1 (Spring-Summer).

Salazar, R., N.P. Louwaars and B. Visser. 2007. "Protecting Farmers' New Varieties: New Approaches to Rights on Collective Innovations in Plant Genetic Resources." *World Development* 35, 9.

Schurman, R., and D. Kelso. 2003. *Engineering Trouble: Biotechnology and its Discontents.* Berkeley, CA: University of California Press.

Shiva, V. 1997. *Biopiracy: The Plunder of Nature and Knowledge.* Boston, MA: South End Press.

Srinivas, K.R. 2006. "Intellectual Property Rights and Bio Commons: Open Source and Beyond." *International Social Science Journal* 58, 319.

Stallman, R. 2002. *Free Software, Free Society.* Boston: Free Software Foundation.

Weber, S. 2004. *The Success of Open Source.* Cambridge, MA: Harvard University Press.

Wright, B.D. 1998. "Public Germplasm Development at a Crossroads: Biotechnology and Intellectual Property." *California Agriculture* 52, 6.

Zimmerer, K. 1996. *Changing Fortunes: Biodiversity and Peasant Livelihood in the Peruvian Andes.* Berkeley, CA: University of California Press.

12

Food Sovereignty in Movement
Addressing the Triple Crisis

Philip McMichael

The food sovereignty movement has emerged as a creative alternative to privatized conceptions of "food security," represented in the "feed the world" mantra of the development establishment. The limits of this mantra have been revealed in the so-called "world food crisis" and the ease with which feeding the world has been replaced by the rush into biofuels for alternative energy. Just as agriculture becomes indistinguishable from energy production, the food sovereignty movement sharpens its claims to "feed the world and cool the planet."[1] In other words, food sovereignty offers a reality check at this moment in time. While small producers (peasants, farmers, fisherfolk, pastoralists, forest dwellers) may not literally be the only stewards of sustainability, it is their model of stewardship that is of value in this global emergency (food, energy and climate crises). This model, in all its diversity, represents a political and ecological counterpart to industrial agriculture.

"Food sovereignty" emerged in the mid 1990s as a strategic challenge to the hegemonic neoliberal concept of food security. The challenge involved a direct politicization of the claims of the WTO and free trade agreements to maximize food security through trade liberalization and agricultural reforms. The food sovereignty movement charged that agricultural liberalization impoverished and displaced small producer populations by allowing "dumping" of artificially cheapened (via massive subsidies) Northern food into Southern markets and by reproducing across the South a new development model of agro-exporting to repay debt. The displacement of local foods by global foods intensifies a colonial pattern of extraction of food resources from South to North. Protections of postcolonial national farm sectors in many countries of the Global South have been dismantled under policies of neoliberal globalization, dedicated to the principle (if not the practice) of universal free trade and the privatization of public services.

While food security depends on a standardized, replicable and ultimately unsustainable agro-monoculture geared to mass producing food for a global market comprising less than 50 percent of the world's population, food sovereignty embodies a richer vision. Not only does it value social access over market supply of food, but decisions about food production (what is produced and how) are grounded in democratic exchanges. The content of the vision comes from the rank and file of the food sovereignty movement itself, a diverse alliance of peasants, farmers, farm workers and indigenous communities in distinct ecological zones and political domains dedicated to a common struggle to promote the rights and practices of small producers (including land rights for the landless) as a condition of democratizing food systems.[2]

These rights and practices serve as the foundation of "agrarian citizenship" (Wittman 2009; Chapter 7 in this volume) in states shifting from competitive to cooperative relations in managing democratic forms of food provisioning, including food exchanges where necessary rather than where profitable. The stated principle of the food sovereignty movement is to protect and promote culturally and environmentally appropriate food production and consumption across a bio-diverse world. While this principle politicizes food security, the realization of food sovereignty in multiple practices occurs at the grassroots level, where *campesinos* "fight for their livelihoods, not for causes" (Holt-Giménez 2006: 182). However, in chronicling the farmer-to-farmer movement that reproduces itself through agro-ecology and horizontal learning networks, Holt-Giménez argues that the strength of the Movimento Campesino a Campesino (MCAC) as a network — "its capacity to generate farmer's agroecological knowledge in a horizontal and decentralized fashion — is also its political weakness as a social movement" (180). This potential hiatus, between local organizations and the transnational movement, is also the subject of Borras's (2008) critique of the Global Campaign for Agrarian Reform (GCAR), which serves as a lens on the organizational difficulties confronting La Vía Campesina in its quest to secure a dialectical integration of "global framing" (diffusion) with local/national campaigns on the land question. This is especially the case now that formalization of hitherto public lands (under neoliberal auspices, as a precursor to privatization) is occurring in regions, such as Africa, where the movement has few local chapters.

As the food sovereignty movement has matured, it has reached a critical organizational juncture. While the central principle of food sovereignty has identified the shortcomings and violence of the neoliberal mantra of food security, the implementation of that principle remains elusive. Certainly the massive dispossession attending neoliberal agrarian and trade reform has so destabilized politics — as clearly demonstrated by the growing number of land protests, food riots and farmer suicides, rising hunger and the fact that the world is becoming a "planet of slums" (Davis 2006) — that neoliberal food security has brought some discredit on itself. But this has not slowed the movement to commodify agricultural relations — as evidenced by the multilateral, political and corporate elite's ignoring of the International Assessment of Agricultural Knowledge, Science and Technology for Development[3] in its repackaging of initiatives to address the world food crisis in Rome in June 2008.

From different angles — Desmarais (2007) (decision-making structure), Holt-Giménez (2006) (externalization capacity), Edelman (2008) (de-peasantization), Borras (2008) (class, ideological and representational disjunctures), Walker Le

Mons (2008) (blind-spot on Chinese land seizures) and Malseed (2008) (regional absences) — all identify organizational weaknesses in the food sovereignty movement. Arguably these issues are part of what might be called "growing pains" (even for Edelman's contrary argument that the Central American *campesino* of a decade ago has disappeared). Borras details how the GCAR has reframed the meaning of agrarian reform, in discourses of rights and responsibilities, contributing to the food sovereignty movement's reframing of global possibility (McMichael 2008). I argue here that the food sovereignty movement faces a historic threshold through which it has the potential to reconstitute its impact practically. This involves seizing the moment by revaluing the viability of local economies with their own political, ecological and social integrity, as bases from which to develop solutions to the combined energy, food and climate crises.

The food sovereignty movement's Terra Preta Declaration at the FAO world food summit in Rome, June 2008, stated:

> The serious and urgent food and climate crises are being used by political and economic elites as opportunities to entrench corporate control of world agriculture and the ecological commons. At a time when chronic hunger, dispossession of food providers and workers, commodity and land speculation, and global warming are on the rise, governments, multilateral agencies and financial institutions are offering proposals that will only deepen these crisis through more dangerous versions of policies that originally triggered the current situation.... Small-scale food producers are feeding the planet, and we demand respect and support to continue. Only food sovereignty can offer long-term, sustainable, equitable and just solutions to the urgent food and climate crises.

This and related declarations propose that the smallholder represents one possible solution to ecological catastrophe and invoke public responsibility for protecting such populations from global market monopolies and dispossession — in the name of rights to food, including grain reserves and regional distribution systems, as well as to work, to produce and to steward the land. The re-embedding of agrifood systems, then, involves a simultaneous embedding of public institutions in local social reproduction — reversing the historical process of state centralization. At the grassroots level, reproducing local knowledges as the foundation for experimental farming is a precondition for the democratic adoption of new technologies appropriate to local conditions and climatic changes as well as for consolidating local and regional food markets. In short, food sovereignty in theory and practice represents a political, ecological and cultural alternative to a "high modernist" corporate agriculture premised on standardized inputs and outputs and serving a minority of the world's population.

Food sovereignty intervenes against the imposition of "agriculture without farmers." Production of "food from nowhere" (Bové and Dufour 2001) accentuates hunger among, and dispossesses, rural majorities across the Global South, compromising states' civic role. In this context the food sovereignty movement advocates an agrarian citizenship dedicated to re-territorializing states through revitalizing local food ecologies under the stewardship of the peasant way: "The government should introduce policies to restore the economic condition of small farmers by providing fair allocation of these production resources to farmers, *recognizing their rights as producers of society*, and recognizing community rights in managing local resources" (La Vía Campesina 2005: 31, emphasis added).

Agrarian citizenship is a tactic appealing to the authority of the state to protect farmers, as guardians of the commons. It is a strategic intervention in the politics of development insofar as it advocates for peasant-farmer rights to *initiate* social reproduction of the economic and ecological foundations of society. In other words, the demand is for rights to be exercised as a means to a social end. Whereas the UN Universal Declaration of Human Rights (1948) defined citizenship as an individual right, or as the "realization, through national effort and international cooperation and in accordance with the organization and resources of each State, of the economic, social, and cultural rights indispensable for his dignity and the free development of his personality" (quoted in Clarke and Barlow 1997: 9), the agrarians view rights as a "means to mobilizing" social relations (Patel 2007: 88). In this mobilization, food is at the centre of the social equation, and this is

> a call for a mass re-politicization of food politics, through a call for people to figure out for themselves what they want the right to food to mean in their communities, bearing in mind the community's needs, climate, geography, food preferences, social mix, and history.... More important, though, is the building of a sustainable and widespread process of democracy that can provide political direction to the appropriate level of government required to see implementation through to completion. (Patel 2007: 91)

The principle of food sovereignty thereby seeks to rebuild small-scale, diverse agricultural systems to resolve modern problems of environmental degradation and provisioning of healthy food to local populations. Drawing on the experimental science of peasant agriculture, under conditions of political mobilization, it is a thoroughly modern response to the current neoliberal conjuncture, which lacks sustainable solutions. This chapter suggests that the food sovereignty movement is well positioned to bring its knowledges of biodiverse and sustainable farming into relation to the ecological challenges facing the world as a viable alternative to unsustainable monocultures of an energy-intensive industrial agriculture, which bears significant responsibility for the triple crisis: displacing local food production for almost 50 percent of humanity, deepening fossil-fuel dependency in an age of "peak oil" and generating roughly a quarter of greenhouse gas emissions (McMichael et al. 2007).

Food Sovereignty

The concept of food sovereignty itself is a hybrid, in large part because it must simultaneously address immediate needs and posit real alternatives. As such, it has both formal and substantive meanings. The formal meaning invokes the right of nations to protect domestic food production and has served to distinguish food sovereignty from food security, based as it is on inequitable neoliberal trade, aid and investment policies. The substantive meaning focuses on the impact on the small producer of these relations and policies, which convert agriculture to export agri-business. In a characteristic press release, La Vía Campesina observes that "small farmers around the world, men and women ... defend the right of countries to protect their domestic markets, to support sustainable family farmers, and to market food in the countries where it is produced" (July 8, 2008). Advocating national food production systems supplied by smallholders is premised on their light ecological footprint and on the claim that they "are currently producing the very large majority of the world's food" (La Vía Campesina 2008b).[4]

The modernity of the food sovereignty movement, then, is not only because it addresses the deficiencies of neoliberal food security but also because it offers the means by which its methods of sustainable food production can address the food, energy and climate crises simultaneously. That is, whether it is the International Planning Committee for Food Sovereignty (IPC) lobbying multilateral and government agencies and reconstructing public discourse, the Brazilian MST working to reverse the displacement of people from countryside to the urban fringe or the *campesino-a-campesino* (farmer-to-farmer) movement sharing sustainable agricultural knowledges and withstanding "both socialist and capitalist versions of progress" (Holt-Giménez 2006: xii), the food sovereignty movement directly addresses the constraints of the moment. These constraints — food insecurities, fossil-fuel dependence and global warming — derive from the long-term assault by a corporate food regime on small producers, whose methods provide the basis of food sovereignty and a key to sustainability.

The European colonization project paralleled an enclosure movement in western Europe, and, through slavery, the occupation and commercialization of agriculture and colonial intervention instigated a series of cycles of incorporation and dispossession of small producers. The subordination of the countryside to capital came to be represented as modernization, where peasantries were considered "backward" and "poor"[5] and, more significantly, increasingly obsolete in the modern world. "Development," then, became a narrative whereby disappearing peasants would serve as labour for industrial expansion, and nutrient cycles (in and through soil and water) would be replaced by rationalized technological systems reducing agriculture to "an input-output process that has a beginning and an end" (Duncan 1996: 123).

In relation to this dynamic, David Harvey distinguishes between contemporary ("anti-globalization") resistances as movements against "accumulation by dispossession" (including peasant mobilization insofar as it resists displacement, declining public supports for small farming and threats to environments, knowledges and cultures) and movements around "expanded reproduction," where the "exploitation of wage labour and conditions defining the social wage are the central issues" (2005: 203). He argues that finding "the organic link between these different movements is an urgent theoretical and practical task" (203). Arguably, it is La Vía Campesina that makes the link between the accelerated circulation of food and the displacement and circulation of people, conditioning the global supply of wage labour and the social wage. That is, the peasant movement, located at the centre of the contemporary crisis of capitalism, is in a position to articulate a comprehensive critique of this development model.

The food sovereignty movement has emerged as an expression of, and potential solution to, the contradictions of agro-industrialization (e.g., food distribution inequity, monoculture, population redundancy, environmental degradation, fossil fuel dependence). While these contradictions represent modern problems without modern solutions (de Sousa Santos 2002), the food sovereignty movement's intervention reframes the agrarian question: namely, under what conditions can food systems respect small producers, environments, ecological knowledges and cuisines? As above, the struggle is a process — first, politicizing neoliberal food security in the name of sovereignty and social rights, and second, developing methods of production and circulation of food that are cooperative, equitable and sustainable. In these senses, food sovereignty's alternative modernity is in restoring agriculture from its historical subordination (and ecological abstraction) to the centre of humanity's relationship with nature (Duncan 1996).[6]

As such, food sovereignty must evolve with circumstances. Since the world food crisis has etched itself into public discourse, with rising prices and food riots, governments have scrambled to shore up their legitimacy by intervening in food markets to increase availability and stabilize prices. Under these conditions, La Vía Campesina has stepped up its appeal to rebuild local and national food systems. The substantive core of food sovereignty is paradigmatic — as the IPC claimed just prior to the June 2008 FAO Food Crisis Summit in Rome:

> Food production and consumption are fundamentally based upon local considerations. The answer to current and future food crises is only possible with a *paradigm-shift toward comprehensive food sovereignty*. Small-scale farmers, pastoralists, fisherfolk, indigenous peoples and others have defined a food system based on the human right to adequate food and food production policies that increase democracy in localized food systems and ensure maximization of sustainable natural resource use. (emphasis added)

In other words, food sovereignty is not simply about revaluing farming as a national good (as opposed to the agro-export model). It also includes an ethic of democratization as the basis of a sustainable agroecological approach to food provisioning in a time of climate and energy emergency. The ethical dimension is critical here, as it has both moral and practical application. Morally, by what right does corporate agriculture displace people, knowledge and culture and induce hunger and malnutrition in the name of an abstraction (the market)? Henry Saragih, member of the International Coordinating Committee of La Vía Campesina, noted: "By reducing the meaning of food to a commodity only those who have money will be able to have access to food" (La Vía Campesina 2008a). And Jean Ziegler, Special Rapporteur to the United Nations, claimed in October 2007 that biofuels are a "crime against humanity," insofar as they displace food production directly. Arguably, this moral issue extends to the commodification of food, as fuel, to sustain carbon-intensive consumerism.

The practical ethic concerns the ecological contributions of small farming: not simply the low-carbon lifestyle of small producers but also in reducing the human ecological stresses of de-peasantization and the build up of slumdwellers. Beyond the moral issue of political and cultural rights to long-standing livelihoods, considerable research and practice supports the claim that small farms are more productive than large mono-cropping agri-business estates (Pretty et al. 2006; Miguel Altieri 2008; Chapter 9 in this volume). Given the conjunction of food, energy and climate crises, the modernist prejudice against smallholder farming, fishing and pastoralism needs reversing. We need to recognize peasant adaptability and resilience to ecological changes under conditions of global warming.[7]

Adaptation

In the context of global warming, the development industry is adopting its own version of adaptation via "climate proofing." A 2008 policy brief for the Commission on Climate Change and Development argues that adaptation, deemed a necessary and "moral responsibility" by the North towards the South is "often similar to, and sometimes indistinguishable from, development." In other words, adaptation (insurance schemes, crop rotations, irrigation systems, drought-resistant seeds, sea defences) reproduces conventional development practices: "Development agencies and NGOs can use their decades of experience in poverty alleviation and sustainable development to assist the poorest countries to meet the adaptation challenge" (Klein et al. 2008: 2). Mainstreaming adaptation "includes 'climate proofing,' i.e., the protection of existing ODA [official development assistance] projects and programs" and ensuring "that future development plans and programs are actively designed to reduce the vulnerability to climate change" (Schaar 2008: 1).

Climate proofing represents a new profit frontier, with agro-chemical and biotechnology firms like BASF, Monsanto, Bayer, Syngenta and Dupont filing over 500 patent documents on so-called "climate ready" genes. Gene patents threaten farmer sovereignty and shift resources away from farmer-based strategies for climate change survival and adaptation:

After decades of seed industry mergers and acquisitions, accompanied by a steady decline in public sector plant breeding, the top 10 seed companies control 57 percent of the global seed market. As climate crisis deepens, there is a danger that governments will require farmers to adopt prescribed biotech traits that are deemed essential adaptation measures. (ETC Group 2008)

These measures are emphasized over documented local initiatives based on adaptive practices, developed largely by women (see, e.g., ActionAid 2007; Lobe 2007; and DDS Community Media Trust et al. 2008). A spokesperson for Monsanto, in a strategic alliance with BASF and the Gates Foundation to develop drought-resistant corn, declared: "I think everyone recognizes that the old traditional ways just aren't able to address these new challenges. The problems in Africa are pretty severe" (quoted in Weiss 2008: 4).

For Africa, research challenges the trope of "traditional ways" and questions typical assumptions that desertification and land degradation result from local mismanagement and misuse of land, rather suggesting that land may have been degraded prior to farming or grazing (Lim 2008). Instead of assuming a linear or path-dependent course of environmental worsening, this "new ecology" approach incorporates variable-temporal patterns in the health of the land, noting that "nature is in a state of continuous change" and "human livelihood adaptations in these environments are very specialized." Lim observes that along the edge of the Sahara, in Nigeria, Niger, Senegal, Burkina Faso and Kenya, integrated farming, mixed cropping and traditional soil and water conservation methods have increased food production substantially ahead of population growth. Further, in eastern Burkina Faso, despite declining rainfall and rising population over the last half century, hoe cultivation has not degraded the land, and food productivity has been maintained and indeed improved for many crops, such as sorghum, millet and groundnuts. He notes: "High local population densities, far from being a liability, are actually essential for providing the necessary labour to work the land, dig terraces and collect water in ponds for irrigation, and to control weeds, tend fields, feed animals and spread manure" (Lim 2008).

Lim observes that population density allows the development of cooperative farming practices and that locals are invested in sustaining social networks, "such as land networks, labour networks, women's natal networks, cattle networks, technology networks and cash networks." In Southern Niger and in Kenya, where droughts have taken an environmental toll, farmers have been able to reverse desertification. Lim concludes: "This is no high-tech breakthrough, nor a result of western aid programs. A major reason for the overestimation of land degradation is the underestimation of local farmers' abilities."

Niger features a success story in the Tahoua region, which has "reclaimed hundreds of hectares of degraded land" and where the president of a women's group remarked on hearing reports claiming degradation of the Sahel, "these experts have never visited us" (Reij 2006). Reij claims that economists underestimate environmental regeneration in focusing their cost-benefit calculations on crop yields, as a surrogate for degradation. Conversely, following environmental and political crises in the 1980s and 1990s respectively, 250,000 hectares of strongly degraded land have been rehabilitated since the mid 1980s: "Dry season cultivation has been expanded considerably and according to FAO statistics Niger produced in 1980 100,000 tons of dry onions, but 270,000 tons in 2004, which was a drought year" (Reij 2006). Further, where farmers have rehabilitated degraded land, household food security has improved. Overall, farmers in high population parts of Niger have succeeded on at least three million hectares in managing natural regeneration, which is unique for the Sahel.

Discounting local farmer knowledges and practices is endemic to an episteme that misconstrues peasant frugality as traditional and incapable of addressing climatic change effects. By labelling them traditional,[8] peasants' resilient practices are not recognized for their intrinsic adaptive qualities. Food sovereignty surely means, in its formal sense, documenting and disseminating these practices as a method of confirming the ability of small producers to, as it were, "feed the world and cool the planet." At the Terra Preta Forum (Rome 2008) a representative from the Network of Peasant Organizations and Agricultural Producers in West Africa (ROPPA by its French acronym) reported that flexible seed selection by farming women in West African has managed recurring drought and that these practices and outcomes are being documented. Similarly, ActionAid's report, *We Know What We Need: South Asian Women Speak Out on Climate Change Adaptation,* documents how women farmers (in the Ganges basin bordering Nepal, India and Bangladesh) manage livelihoods under conditions of erratic monsoon patterns, concluding that their evidence "proves that women in poor areas have started to adapt to a changing climate and can clearly articulate what they need to secure and sustain their livelihoods more effectively" (2007: 4).

Generally, development agencies are not geared to support peasants' and small farmers' ingenuity. In another ecological zone, the drylands of the Deccan Plateau, where a variety of rainfed crops, including sorghum, millets, pulses and oilseeds, grow, the "symbiotic relationship between these crops provides solutions to a wide range of problems faced in today's Indian agriculture such as management of soils and their fertility, pest control as well as minimizing risk and uncertainty" (DDS Community Media Trust et al. 2008: 35). While such biodiversity may allow farm communities to manage climatic conditions in tenuous environments,

> the many values of uncultivated biodiversity used by people for food, fodder and medicine have generally been unseen and neglected by the official discourse.... The number of uncultivated foods that are harvested in Medak district (Andhra Pradesh) greatly exceeds the number of cultivated species. Some 80 species of uncultivated leafy greens are locally used as foods and

many dozens more species of uncultivated plants including roots, tubers and fruits. This vast array of "wild" leafy greens, berries and fruits are sources of many nutrients…. Most of them are rich sources of calcium, iron, carotene, vitamin C, riboflavin and folic acid. Therefore they are a boon to pregnant and nursing women as well as to young children. Since they come at no monetary cost at all, they are a blessing for the poor. Dalits know it and have woven these uncultivated foods into their food system. (DDS Community Media Trust et al. 2008: 35)

Non-monetized resources such as wild plants and exchange networks do not count in the view of the development industry.

Seed Politics

In her study of seed diversity in the dry Deccan Plateau of South India, Carine Pionetti (2005) documented the value of women's seed work in forming a "localized seed economy" through seed exchanges, which have ecological, economic, social and cultural significance. In contrast to the commercialization of seeds — meaning their monopolization under patents favoured by the development industry's quest to feed the world:

> The continuous exchange of seeds for local crop varieties circulates genetic resources from one field to another within a village territory and beyond. The dynamic management of genetic resources enhances the stability of traditional agrosystems, increases the adaptation potential of local crops to evolving environmental conditions and limits the risk of genetic erosion. Seed transactions also help ensure that land is not left fallow for lack of seeds, thus avoiding soil erosion and increasing the soil's organic matter content and water retention capacity. (Pionetti 2005: 154)

Pionetti's point is that seed saving minimizes risk, increases crop diversity and nutrition, provides "self-reliance and bargaining power within the household" (xiv) for women, allows women to select seeds to meet specific individual, environmental and climatic needs, allows planting at appropriate times and provides assets (seeds constitute a currency, particularly among women with few resources). Seeds constitute the security of a "knowledge commons" (Holt-Giménez 2006: 97). Once farmers buy into the agro-industrial system, they are "locked into a production chain where the choices of inputs and the use of the harvest are predetermined by agro-chemical and food-processing firms" (Pionetti 2005: xv), and farmers "become dependent on a network of technical information generated by specialists (agricultural scientists, chemists, genetic engineers, nutritionists …) and transferred to farming communities by agricultural extension workers or technicians" (166).

Seed commercialization is the principal legacy of the Green Revolution.

Whereas the first Green Revolution was a publicly funded initiative to provide staple grain foods but simultaneously displacing small farmers unable to "buy in" and dryland farmers in India undercut by rice and wheat circulated through the Public Distribution System),[9] the second Green Revolution (associated with the neoliberal food security project) is a privately funded initiative for high-value foods (meats, fruits and vegetables) for relatively affluent global consumers. It is within this framework that the commercialization of seed to develop African agricultural resources is anticipated by the Rockefeller/Gates Alliance for a Green Revolution in Africa (AGRA). The AGRA, one of the key projects identified at the Rome Summit (2008) by the development industry, is premised on expansion of the "outgrower" model, featured in the World Bank's plan for tackling the food crisis, including "more investment in agri-business so that we can tap the private sector's ability to work across the value chain" (Zoellick 2008). The "value chain" is to be anchored in genetically modified crops using "no-till farming" (a less soil-disturbing technology that nevertheless delivers GM seeds) as part of a "green" agenda (GRAIN 2009).

The green agenda represents a corporate attempt to compensate for the deep-ening of the separation of producers from natural cycles. Separation is not simply the elimination of nutrient recycling in agriculture — based in "crop sequencing, crop rotation, fallowing, weeding, selective clearing, intercropping, appropriate crop and landrace selection, adapted plant spacing, thinning, mulching, stubble grazing, weeding mounds, paddocking, household refuse application, manure application, weeding mounds, crop residue application and compost pits" (Lim 2008) — but is also inherent in the privatization of seed. Seeds and seed saving are the foundation of food sovereignty (La Vía Campesina 2008a: 38). But the struggle is with governments as much as with corporations, since states are under pressure to adopt "value chain" agriculture. Thus Thai farmers, under pressure from the state, "have transformed their upland sustainable agriculture to cash crops for exports and market oriented and high productive seed varieties are promoted intensively in flat areas to replace native seeds which are suitable to local geo-cultural condi-tions" (La Vía Campesina 2008a: 45). Farmers replace rotational cultivation with commercial agriculture, with seeds supplied by "CP, Monsanto and other seeds dealers selling seeds coming from Japan and China" (47).

La Vía Campesina points to the power dynamics involved by stressing that "seed is becoming the object of exploitation of the farmers because it is being controlled and managed by big agri-business companies that have tie-up with Transnational Companies like Dupont, Monsanto, Syngenta, Bayer, Limagrain, Dow and Aventia" (2008a: 48). The Federation of Indonesian Peasants Union (FSPI) reports: "The Indonesian government in its effort to increase rice paddy production because of the threat of a food crisis imported the hybrid seeds and distribute it to the farm-ers, making them more and more dependent on these seeds" (quoted in La Vía Campesina 2008a: 48). In this way the food crisis may deepen centralized rule through privatizing seed markets in order to commandeer food from national hinterlands.

Meanwhile, Karen communities in Northern Thailand jealously conserve native varieties of rice, grains, vegetables and herbs via rotational cultivation, and the Assembly of the Poor uses networks like the Alternative Agricultural Network to collect rice varieties, use agroecology methods and promote household consumption (2008a: 46). In Indonesia, the FSPI leads resistance to government attempts to increase national rice production through hybrid seeds.

The deepening of peasant and farmer dependency on commercial inputs is foreshadowed in the World Bank's *World Development Report* 2008, where the Bank, in an apparent turnaround, advocates "'market-smart' subsidies to stimulate input markets" meaning, presumably, subsidized fertilizer and seeds for "improving" farmers.[10] This market-led agriculture is to be secured with "risk management instruments" including "weather-based index insurance," but with skepticism towards public grain reserves, which countries manage "with very mixed success."[11] For rain-fed areas, the World Bank proposes that one of "agriculture's major success stories in the past two decades is conservation (or zero) tillage" (2007: 16) — as mentioned above, a less soil-disturbing technology that nevertheless delivers GM seeds along with networks of fertilizer dealers, thereby deepening "petro farming" and the control of the chemical and seed companies.

Energy Security

The claim to "feed the world" via improving productivity, in enforcing property rights in seeds and other inputs, brings further "accumulation by dispossession" (Harvey 2003) rather than supporting farmers' markets geared to feeding the working poor, a substantial portion of humanity. It parallels a new enclosure of the commons, driven by a related security concern: energy supply. Governments are now identifying what they consider, mistakenly, as "idle lands" to be commandeered for agrofuels production for valuable energy exports (Cotula et al. 2008: 22–23). Given the controversy surrounding the competition between biofuel and food crops, proposals are emerging for biofuel crops to occupy land deemed marginal (Gaia Foundation et al. 2008: 1). Thus the recent U.K. Gallagher Report notes that current policies are not geared to making use of suitable land for biofuels and that idle or marginal land should be targeted (2008:9). Merging "marginal land" with "abandoned cropland" underlies a number of "bioenergy feasibility studies" that inform policy.

However, much of this land supports the subsistence needs of local rural populations, such as pasturing livestock, fuel, medicine and building material. The Gaia Foundation notes that common land used for generations may not be titled, that its fragility means it is used sparingly over the years and that it is often sacred to communities and may be vital to protecting water sources (2008: 3). In addition to displacing communities to more fragile regions, or peri-urban sites, government-sanctioned land acquisitions in the name of energy security may undermine customary use and ecological practices geared to sustaining landscapes. In India, for example, *jatropha* production targets "waste lands" that sustain millions of people as "commons" and pasturelands. In addition, GRAIN reports:

> Refugees from development projects, displaced persons, jobless labourers and small farmers facing crop failure often rely on these lands as places where they can put their cattle during an emergency. If these lands are enclosed, the lifelines of many already disadvantaged people will be jeopardized. (2008: 8)

While a recent FAO report emphasizes the significance of marginal lands for subsistence functions of the rural poor, primarily women with no property rights, the Gaia Foundation nevertheless notes:

> It is no coincidence that the livelihoods of communities who do not practice intensive agriculture and in particular of pastoralists and women are being ignored in the debate. While deforestation for agrofuels is seen as something to be avoided, the conversion of pasture lands and non-intensively farmed lands in the South is regarded as essential and desirable if bioenergy is to replace a significant amount of fossil fuels in industrial societies. (2008: 4)

Conclusion

The food sovereignty movement confronts a series of new challenges, the most critical flowing from the conjunctural crises of food, energy and climate change. The movement has politicized the neoliberal project of export agriculture, and it is now seizing the initiative to represent smallholding as a solution to the new threshold of crisis faced by the global community. Doing so requires deepening public recognition of the *social* value of producer sovereignty and agroecological practices. To invoke practical conditions for a food sovereignty counter-narrative, the food sovereignty movement is embarking on new alliances to create

> new counter-narratives and counter-images... which can hold their own against the tales, metaphors and rituals binding the networks for whom economic scarcity and the inevitability of wholesale privatization... remain sacred writ. What stories can rural dwellers relate about how they came to build up their forest islands or improve their soils?... What metaphors can tie such narratives together in ways which engender new commitments and mutual responsibility? (Lohmann 2003: 20)

In other words, food sovereignty needs not just a politics of peasant rights, but also a politics of representation (Patel 2006) of smallholder capacity, whereby the salience of peasant agriculture at this time becomes a legitimate and compelling part of public discourse. Peasant alliance counter-narratives are necessary to anchor the claims that, as Marx maintained (and as current history demonstrates):

> The moral of the tale … is that the capitalist system runs counter a rational agriculture, or that a rational agriculture is incompatible with the capitalist system (even if the latter promotes technical development in agriculture) and needs either small farmers working for themselves or the control of the associated producers. (quoted in Foster 2000: 165)

But these counter-narratives need grounding — in informing a public discourse regarding world-historical solutions to the combined crises facing the planet.

In his analysis of the politics of MCAC, Holt-Giménez details the "capillary action" of networks in generating circuits of knowledge and seed exchange, but in regenerating agroecological *practices*, farmer networks were unable to focus on the socioeconomic and political *conditions* for sustainable agriculture (2006: 180). His solution is to introduce dialogue about structural issues into MCAC workshops and networks, essentially building alliances between farmer networks and NGOs. This supports movements that are repositioning sustainable agriculture as an agroecological condition for sustaining human/e life on this planet. Meanwhile, Edelman suggests:

> [The] difficulty in expanding norms about rights is part of a broader problem of making peasant voices heard in societies where thousands are abandoning the countryside every day and where powerful elites and policymakers no longer view agriculture as the motor force of economic development. (2008: 251)

His point is that the Central American *campesino* today is a pluriactive off-farm operator with remittances and urban connections and is even less likely to form alliances with capital-intensive farmers producing non-traditional exports. This may be so (and pronounced in Central America), but it is also the case that in the present conjuncture, while the development industry is refocusing on "agriculture for development," the food sovereignty movement and its allies are reframing both "development" and "agriculture" in relation to a politicized understanding of sustainability.

Arguably, there are ways to make more explicit the opportunity and potential of smallholders (and, by extension, reconstituted smallholding agricultures) to "feed the world and cool the planet." The IAASTD Report (2008) is a significant step in the right direction, as are the unfolding second (public, if not private) thoughts about the contradictions of agrofuels. These are opportunities the food sovereignty movement needs to seize and nurture, with mounting evidence of the appropriateness of a rational, rather than a rationalized, agriculture that is entirely modern in addressing the catastrophe in motion.

Notes

The author is grateful to an anonymous reviewer and the editors for helpful suggestions in refining the presentation of this chapter.

1.   See the Terra Preta declaration from the Rome summit (2008), which this author had a hand in preparing. Available at <viacampesina.org/main_en>.

2.   In relation to this vision, Borras (2008: 262) notes that La Vía Campesina and the International NGO/CSO Planning Committee for Food Sovereignty (IPC) have evolved a more comprehensive Global Campaign for Agrarian Reform (GCAR), consolidating a "human-rights-based approach," and an alternative vision, beyond a critique of Bank-initiated "market led agrarian reform."

3.   The IAASTD Report advocates a *multifunctional* role for agriculture in reducing poverty and social/gender inequality as well as reversing environmental degradation and global warming, noting that *extant* agriculture "generates large environmental externalities, many of which derive from failure of markets to value environmental and social harm and provide incentives for sustainability" (2008: 20).

4.   Wayne Roberts notes peasants constitute "over a third of the world's population and two-thirds of the world's food producers" (2008), while Miguel Altieri (2008) claims that small farmers (2 hectares and less) produce the majority of staple crops for urban and rural inhabitants across the world.

5.   Jeffrey Sachs recycles this "truth" in *The End of Poverty*: "The move from universal poverty to varying degrees of prosperity has happened rapidly in the span of human history. Two hundred years ago… [j]ust about everybody was poor, with the exception of a very small minority of rulers and large landowners" (2005: 26).

6.   Reflecting on modern (social) scientific thought's deficiencies with respect to planetary ecology (and agriculture's place), C.F. Duncan (2007) notes that "instead of rethinking our possible role on the planet, instead of qualifying or revising our anthropocentric habits, we have shamelessly used the decline of theology relative to science as an excuse to elevate our own importance further. Logically we should have replaced theology with ecology, before enlarging the parameters of our behaviour by the heavy use of fossil fuels."

7.   Surveys of the devastation of Hurricane Mitch in Central America show that "many small farmers cope and even prepare for climate change, minimizing crop failure through increased use of drought tolerant local varieties, water harvesting, mixed cropping, opportunistic weeding, agro-forestry and a series of other traditional techniques" (Altieri 2008). See also Holt-Giménez (2006: 196–97), ActionAid (2007) and DDS Community Media Trust et al. (2008).

8.   Whether poor by Northern standards (footnote 5, where Sachs equates frugality with poverty), or not.

9.   The DDS Community Media Trust et al. note that the redistribution of Green Revolution cereals to food deficit regions "have displaced dryland farmers from their agriculture, undermined the nutrition and food security of poorer communities, marginalized women farmers, robbed farmers of markets for their cereals (sorghum and millets) and eroded biodiversity important for food, farming and ecosystem resilience" (2008: 33).

10.   World Development Report Policy Brief: "More and Better Investment in Agriculture."

11.   World Development Report Policy Brief: "Agenda for Agriculture-Based Countries

of Sub-Saharan Africa." It is common knowledge that the multilateral agencies dismantled grain reserves. As La Vía Campesina noted: "National food reserves have been privatized and are now run like transnational companies. They act as speculators instead of protecting farmers and consumers" (2008c).

References

ActionAid. 2007. "We Know What We Need. South Asian Women Speak Out on Climate Change Adaptation." Institute of Development Studies. Available at <http://www.reliefweb.int/rw/lib.nsf/db900SID/PANA-79DK4C?OpenDocument>.

Altieri, M. 2008. "Small Farms as a Planetary Ecological Asset: Five Key Reasons Why We Should Support the Revitalization of Small Farms in the Global South." Food First. Available at <foodfirst.org/en/node/2115>.

Borras, S.M. Jr. 2008. "La Vía Campesina and its Global Campaign for Agrarian Reform." *Journal of Agrarian Change* 8, 2/3.

Bové, J., and F. Dufour. 2001. *The World Is Not For Sale*. London: Verso.

Clarke, T., and M. Barlow. 1997. MAI: *The Multilateral Agreement on Investment and the Threat to Canadian Sovereignty*. Toronto: Stoddart.

Cotula, L., N. Dyer and S. Vermeulen. 2008. "Fuelling Exclusion? The Biofuels Boom and Poor People's Access to Land." International Institute for Environment and Development (IIED) and Food and Agricultural Organization (FAO). Available at <http://www.iied.org/pubs/pdfs/12551IIED.pdf>.

Davis, M. 2006. *Planet of Slums*. London and New York: Verso.

DDS Community Media Trust, P.V. Satheesh and M. Pimbert. 2008. *Affirming Life and Diversity. Rural Images and Voices on Food Sovereignty in South India*. London: International Institute for Environment and Development/IIED and the Deccan Development Society.

Desmarais, A.A. 2007. *La Vía Campesina: Globalization and the Power of Peasants*. Halifax: Fernwood Books.

Duncan, C. 1996. *The Centrality of Agriculture. Between Humankind and the Rest of Nature*. Montreal: McGill Queens University Press.

_____. 2007. "The Practical Equivalent of War? Or, Using Rapid Massive Climate Change to Ease the Great Transition Towards a New Sustainable Anthropocentrism." Available at <globetrotter.berkeley.edu/GreenGovernance/Rapid%20Climate%20Change.pdf>.

Edelman, M. 2008. "Transnational Organizing in Agrarian Central America: Histories, Challenges, Prospects." *Journal of Agrarian Change* 8, 2/3.

ETC Group. 2008. "Patenting the 'Climate Genes'... and Capturing the Climate Agenda." ETC Group Communique 99 (May/June). Available at <etcgroup.org>.

Foster, J.B. 2000. *Marx's Ecology: Materialism and Nature*. New York: Monthly Review Press.

Gaia Foundation, Biofuelwatch, the African Biodiversity Network, S. La Selva, Watch Indonesia and EcoNexus. 2008. "Agrofuels and the Myth of the Marginal Lands." September. Available at <gaiafoundation.org/documents/Agrofuels&MarginalMyth.pdf>.

Gallagher, E. 2008. *The Gallagher Review of the Indirect Effects of Biofuels Production*. UK Government: Renewable Fuels Agency.

GRAIN. 2008. "Agrofuels in India, Private Unlimited." *Seedling* April. Available at <grain.org>.

_____. 2009. "Nerica: Another Trap for Small Farmers in Africa." GRAIN Briefing, January. Available at <grain.org/go/nerica>.

Harvey, D. 2003. *The New Imperialism*. Oxford: Oxford University Press.

_____. 2005. *A Brief History of Neoliberalism*. Oxford: Oxford University Press.

Holt-Giménez, E. 2006. *Campesino a Campesino. Voices from Latin America's Farmer to Farmer Movement for Sustainable Agriculture*. San Francisco: Food First Books.

International Assessment of Agricultural Knowledge, Science and Technology for Development (IAASTD). 2008. "Executive Summary of the Synthesis Report." Available at <agassessment.org/docs/SR_Exec_Sum_280508_English.pdfLo>.

Klein, R., T. Banuri, S. Kartha and L. Schipper. 2008. "International Climate Policy." Commission on Climate Change and Development, March. Available at <ccdcommission.org>.

La Vía Campesina. 2005. "Impact of the WTO on Peasants in South East Asia and East Asia." *Jl. Mampang Prapatan* XIV, 5. Available at <viacampesina.org/main_en/index.php?option=com_content&task=blogsection&id=8&itemid=30>.

_____. 2006. "Sovranita Alimentare. Final Declaration: For a New Agrarian Reform Based on Food Sovereignty." March 9. Available at <viacampesina.org/main_en/index.php?option=com_content&task=view&id=180&Itemid=27>.

_____. 2008a. "Seed Heritage of the People for the Good of Humanity." From the Women Seed Forum in South Korea. Available at <viacampesina.net/downloads/PDF/seed_heritage_of_the_people_for_the_good_of_humanity.pdf>.

_____. 2008b. "The G8 Is Using the Food Crisis to Promote Their Free Trade Agenda." Press release. July 9.

_____. 2008c. "A Response to the Global Food Crisis: Sustainable Family Farming can Feed the World." Press release. February 15.

Lim, L.C. 2008. "Sustainable Agriculture Pushing Back the Desert." Institute of Science in Society. Available at <i-sis.org.uk/desertification.php>.

Lobe, K. 2007. "A Green Revolution of Africa: Hope for Hungry Farmers?" Canadian Foodgrains Bank. Amersfoort: ILEIA.

Lohmann, L. 2003. "Re-Imagining the Population Debate." The Corner House, Briefing 28. Available at <thecornerhouse.org>.

Malseed, K. 2008. "Where There is No Movement: Local Resistance and the Potential for Solidarity." *Journal of Agrarian Change* 8, 2/3.

McMichael, A.J., J.W. Powles, C.D. Butler, and R. Uauy. 2007. "Food, Livestock Production, Energy, Climate Change and Health." *The Lancet* 370, 9594.

McMichael, P. 2005. "Global Development and the Corporate Food Regime." In F.H. Buttel and P. McMichael (eds.), *New Directions in the Sociology of Global Development*. Oxford: Elsevier Press.

_____. 2008. "Peasants Make Their Own History, But Not Just as They Please ..." *Journal of Agrarian Change* 8, 2/3.

Orth, M. 2007. *Subsistence Foods to Export Goods: The Impact of an Oil Palm Plantation on Local Food Sovereignty, North Barito, Central Kalimantan, Indonesia*. Sawit Watch. Wageningen: Van Hall Larenstein.

Patel, R. 2006. "International Agrarian Restructuring and the Practical Ethics of Peasant Movement Solidarity." *Journal of Asian and African Studies* 41, 1–2.

_____. 2007. "Transgressing Rights: La Vía Campesina's Call for Food Sovereignty." *Feminist Economics* 13, 1.

Pionetti, C. 2005. *Sowing Autonomy: Gender and Seed Politics in Semi-Arid India*. London: IIED.

Pretty, J.N., A.D. Noble, D. Bossio, J. Dixon, R.E. Hine, F.W.T. Penning de Vries, and J.I.L. Morison. 2006. "Resource Conserving Agriculture Increases Yields in Developing Countries." *Environmental Science & Technology* 40, 4.

Reij, C. 2006. "More Success Stories in Africa's Drylands than Often Assumed." ROPPA (The Network of Peasant Organizations and Producers in West Africa). Available at <roppa.info?IMG/pdf/More_success_stories-in_Africa-Reij_Chris.pdf>.

Roberts, W. 2008. *The No-Nonsense Guide to World Food*. Oxford: New Internationalist.

Sachs, J. 2005. *The End of Poverty: Economic Possibilities for Our Time*. New York: Penguin.

Santos, B. de Sousa. 2002. *Towards a New Legal Common Sense*. London: Butterworth.

Schaar, J. 2008. "Overview of Adaption Mainstreaming Initiatives." Commission on Climate Change and Development, March. Available at <ccdcommission.org>.

Terra Preta. 2008. "Civil Society Declaration of the Terra Preta Forum." Terra Preta Forum of the Food Crisis, Climate Change, Agro-Fuels and Food Sovereignty. Rome, June 6. Available at <www.viacampesinba.org/main_en>.

Walker, K. Le Mons. 2008. "From Covert to Overt: Everyday Peasant Politics in China and the Implications for Transnational Agrarian Movements." *Journal of Agrarian Change* 8, 2/3.

Weiss, R. 2008. "Firms Seek Patents on 'Climate Ready' Altered Crops." *Washington Post*, May 13.

Wittman, H. 2009. "Reframing Agrarian Citizenship: Land, Life and Power in Brazil." *Journal of Rural Studies* 25.

World Bank. 2007. *World Development Report 2008*. Washington, DC: World Bank.

Zoellick, R. 2008. "A 10-Point Plan for Tackling the Food Crisis." *Financial Times*, May 29.

13

What Does Food Sovereignty Look Like?

Raj Patel

Hannah Arendt observes that the first right, above all others, is the right to have rights. In many ways, La Vía Campesina's call for food sovereignty is precisely about invoking a right to have rights over food. But it's unclear quite how to cash out these ideas. This chapter looks at some of the difficulties involved in understanding the idea of communities', peoples' and states' rights over food policy, particularly as they concern tensions between different geographies of citizenship. Such tensions are likely to arise not only between producers and consumers but within the bloc of "small farmers" itself, along axes of power that range from patriarchy to feudalism. This chapter suggests that resolution to this conflict is likely to come not through a specific set of rights but rather through the principles necessary to create the preconditions for food sovereignty.

There is, among those who use the term, a strong sense that while "food sovereignty" might be hard to define, it is the sort of thing one knows when one sees it. This is a little unsatisfactory. There is a need to put a little more flesh on the concept's bones, beyond the widely agreed notion that it isn't what we have at the moment. It's worth looking first at the genealogy of the term and observing in particular that it is itself a precondition of food security. That food sovereignty should be the *sine qua non* of food security suggests the following question: what are the preconditions for food sovereignty? Answering this question offers, I argue, a helpful way to generate some substantive political principles with which to recognize food sovereignty before seeing it.

The Etymology of "Food Sovereignty"

It is, admittedly, the first instinct of an uninspired scholar to head towards definitions. I have, far more frequently than I'd care to remember, plundered the Oxford English Dictionary for an authoritative statement of terms against which I then tilted. The problem with food sovereignty is, however, a reverse one. Food sovereignty is, if anything, over defined. There are so many versions of the concept, it's hard to know exactly what it means. The proliferation of statements is, however, woven into the fabric of food sovereignty of necessity. Since food sovereignty is a call for peoples' rights to shape and craft food policy, it can hardly be surprising that this right is used to explore and expand the covering political philosophy. Yet this exploration has sometimes muddled and masked some difficult contradictions within the notion of food sovereignty.

Before going into those definitions, it's worth contrasting food sovereignty with the concept against which it has traditionally been ranged: food security. The FAO has done a fine job of tracking the evolution of "food security" (see FAO 2003), but it is useful to be reminded that the first official definition, in 1974, was: "the availability at all times of adequate world food supplies of basic foodstuffs to sustain a steady expansion of food consumption and to offset fluctuations in production and prices" (United Nations 1975, cited in FAO 2003). The utility of the term then derived from its political economic time, in the midst of the Sahelian famine, at the zenith of demands for a new international economic order and the peak of Third Worldist power, which had already succeeded in establishing the United Nations Conference on Trade and Development as a bastion of commodity price stabilization (Rajagopal 2000). In such a context, when states were the sole authors of the definition and when there was a technocratic faith in the ability of states to redistribute resources, if the resources could only be made available, it made sense to talk about sufficient world supplies and for the primary concern of the term's authors to lie in price stabilization.

Compare the language and priorities reflected in the early 1970s definition to this more recent one: "Food security [is] a situation that exists when all people, at all times, have physical, social and economic access to sufficient, safe and nutritious food that meets their dietary needs and food preferences for an active and healthy life" (FAO 2001, cited in FAO 2003). The source for this definition was *The State of Food Insecurity 2001*, and herein lies some of the tale in the widening gyre of "food security." The definition in 2001 was altogether more sweeping. While it marked the success of activists, NGOs and the policymaking community both to enlarge the community of authors of such statements to include non-state actors and to shift the discussion away from production issues towards broader social concerns, it was also an intervention in a very different world and series of debates. No longer was there a Non-Aligned Movement. Nor was there, at least in the world of state level diplomacy, the possibility of an alternative to U.S.-style neoliberal capitalism. It was an intervention at a time when neoliberal triumphalism could be seen in the break away from a commitment to the full meeting of human rights, to the watered down Millennium Development Goals (which allowed and sanctioned the continuing violation of the right to food). The early 2000s was also a time when the institutions originally created to fight hunger, such as the Food and Agriculture Organization of the United Nations, were looking increasingly irrelevant and cosmetic in decision-making around hunger policy. The expansion of the definition of food security in 2001, in other words, was both a cause and consequence of its increasing irrelevance as a guiding concept in the shaping of international food production and consumption priorities.

While harsh, this assessment is not unreasonable. The terms on which food is, or isn't, made available by the international community have been taken away from institutions that might be oriented by concerns of food security and given to the market, which is guided by an altogether different calculus. It is, then, possible to tell a coherent story of the evolution of "food security" by using the term as a mirror of international political economy.

But that story isn't one in which capital is dominant: "food security" moved from being simply about producing and distributing food to a whole nexus of concerns around nutrition, social control and public health. In no small part, that broadening was a direct result of the leadership taken by La Vía Campesina to introduce at the World Food Summit in 1996 the idea of food sovereignty, a term that was specifically intended as a foil to the prevailing notions of food security, which, almost studiously, avoided discussing the social control of the food system. As far as the most recent definition of food security is concerned, it is entirely possible for people to be food secure under a dictatorship. From a state perspective, the absence of specification about how food security should come about was diplomatic good sense: to introduce language that committed member states to particular internal political arrangements would have made the task of agreeing to a definition considerably more difficult. But, having been at the whip end of structural adjustment and other policies that had the effect of "depeasantizing" rural areas under the banner of increasing food security by increasing efficiency (Araghi 1995), La Vía Campesina's position was that a discussion of internal political arrangements was a necessary part of the substance of food security. Indeed, food sovereignty was declared a logical precondition for the existence of food security:

> Long-term food security depends on those who produce food and care for the natural environment. As the stewards of food producing resources we hold the following principles as the necessary foundation for achieving food security…. Food is a basic human right. This right can only be realized in a system where food sovereignty is guaranteed. *Food sovereignty is the right of each nation to maintain and develop its own capacity to produce its basic foods respecting cultural and productive diversity.* We have the right to produce our own food in our own territory. Food sovereignty is a precondition to genuine food security. (La Vía Campesina 1996, emphasis added)

To raise questions about the context of food security, and therefore to pose questions about the relations of power that characterize decisions about how food security should be attained, was a shrewd move. The first exposition of food sovereignty recognized right at the beginning that the politics of the food system needed explicitly to feature in the discussion. In the context of an international meeting, at a time of unquestioned U.S. hegemony and given states' reluctance to discuss the means through which food security was to be achieved, it made sense to deploy language to which states had already committed themselves. Thus the language of food sovereignty inserts itself into international discourse by making claims on rights and democracy, the cornerstones of liberal governance.

Big Tents & Rights Talk

The outlines of food sovereignty have been well rehearsed elsewhere (see earlier chapters of this book, as well as Rosset 2003; Windfuhr and Jonsén 2005; McMichael 2008). The common denominator in these accounts is the notion that the politics of food sovereignty is something that requires direct democratic participation, an end to the dumping of food, the wider use of food as a instrument of policy, comprehensive agrarian reform and respect for life, seed and land. But as the exponents of food sovereignty, myself included, have begun to explore what this might mean, things have started to look increasingly odd. The term has changed over time, just like food security, but while it's possible to write an account of the evolution of food security with reference to changing international politics, a similar story is much harder to make coherent in reference to the changes with food sovereignty. From the core of the 1996 definition, italicized above, consider this one, written six years later:

> Food sovereignty is the right of peoples to define their own food and agriculture; to protect and regulate domestic agricultural production and trade in order to achieve sustainable development objectives; to determine the extent to which they want to be self reliant; to restrict the dumping of products in their markets; and to provide local fisheries-based communities the priority in managing the use of and the rights to aquatic resources. Food sovereignty does not negate trade, but rather, it promotes the formulation of trade policies and practices that serve the rights of peoples to safe, healthy and ecologically sustainable production. (Peoples Food Sovereignty Network 2002: 1)

It's a cautious definition, talking about a right to define food policy, sensitive to the question of whether trade might belong in a world with food sovereignty. Perhaps most clearly, it's a definition written in committee. The diversity of opinions, positions, issues, and politics bursts through in the text — from the broad need for sustainable development objectives to the specific needs of fishing villages to manage aquatic resources. This is an important strength. Food sovereignty is a big tent, and the definition reflects that very well indeed.

The idea of "big tent" politics is that disparate groups can recognize themselves in the enunciation of a particular program. But at the core of this programme there needs to lie an internally consistent set of ideas.[1] It's a core that has never fully been made explicit, which might explain why in more recent definitions of food sovereignty, increasing levels of inconsistency can be found. Consider this statement, from La Vía Campesina's Nyéléni Declaration:

> Food sovereignty is the right of peoples to healthy and culturally appropriate food produced through ecologically sound and sustainable methods, and their right to define their own food and agriculture systems. It puts those who produce, distribute and consume food at the heart of food systems and policies rather than the demands of markets and corporations. It defends the interests and inclusion of the next generation. It offers a strategy to resist and dismantle the current corporate trade and food regime, and directions for food, farming, pastoral and fisheries systems determined by local producers. Food sovereignty prioritises local and national economies and markets and empowers peasant and family farmer-driven agriculture, artisanal fishing, pastoralist-led grazing, and food production, distribution and consumption based on environmental, social and economic sustainability. Food sovereignty promotes transparent trade that guarantees just income to all peoples and the rights of consumers to control their food and nutrition. It ensures that the rights to use and manage our lands, territories, waters, seeds, livestock and biodiversity are in the hands of those of us who produce food. Food sovereignty implies new social relations free of oppression and inequality between men and women, peoples, racial groups, social classes and generations. (La Vía Campesina 2007: 1)

The contradictions in this definition are a little more fatal. The phrase "those who produce, distribute and consume food" refers, unfortunately, to everyone, including the transnational corporations rejected in the second half of the sentence. There's also a glossing over of one of the key distinctions in agrarian capitalism — that between farm owner and farm worker. To harmonize these two groups' interests is a far less tractable effort than the authors of the declaration might hope. Finally, but perhaps most contradictory, is the emphasis on "new social relations" in the same paragraph as family farming, when the family is one of the oldest factories for patriarchy.

There are, of course, ways to smooth out some of these wrinkles. One might interpret "those who produce, distribute and consume food" as natural rather than legal people. Corporations aren't flesh and blood, and while they might be given equal rights as humans, there are growing calls for the privilege to be revoked (Bakan 2004). Even if one accepts this definitional footwork, we remain with the problem that, even between human producers and consumers in the food system, power is systematically unevenly distributed.

· One way to balance these disparities is through the explicit introduction of rights-based language. Necessarily, to talk of a *right* to shape food policy is to contrast it with a *privilege*. The modern food system has been architected by a handful of privileged people. Food sovereignty insists that this is illegitimate because the design of our social system isn't the privilege of the few but the right of all. By summoning this language, food sovereignty demands that such rights be respected, protected and fulfilled, as evinced through twin obligations of conduct and result (Balakrishnan and Elson 2008). It offers a way of fencing off particular entitlements by setting up systems of duty and obligation.

Hannah Arendt & the Right to Have Rights

Hannah Arendt is perhaps the most appropriate theorist to bring to bear here, not least because in her *Origins of Totalitarianism*, she makes an observation about rights strikingly similar to that motivating food sovereignty:

> People deprived of human rights ... are deprived, not of the right to freedom, but of the right to action, not of the right to think whatever they please, but of the right to opinion .... We become aware of the existence of a right to have rights (and that means to live in a framework where one is judged by one's actions and opinions) and a right to belong to some kind of organized community, only when millions of people emerge who had lost and could not regain these rights because of the new global political situation. (Arendt 1967: 177)

Although referring to European colonialism and its dehumanizing blowback in the metropole in talking about how humans are disarmed and rendered unable to effect change in the world around them, Arendt could have been describing the contemporary context of food politics — well, perhaps with the caveat that the political situation has *never* been favourable to those who produce food; its new global context merely compounds a millennia-old disenfranchisement.

But despite its apparent applicability, the language of rights doesn't come cheap, and it might not be well suited to the idea of food sovereignty. Central to the idea of rights is that a state is ultimately responsible for guaranteeing the rights within its territory, because it is sovereign over it. As I have written elsewhere (Patel 2006), this understanding of the agency required for rights to proceed is something that Jeremy Bentham has put rather directly: "Natural rights is simple nonsense: natural and imprescriptible rights, rhetorical nonsense — nonsense upon stilts" (Bentham 2002: 330). The argument that Bentham makes is simple: rights cannot be summoned out of thin air. For rights to mean anything at all, they need a guarantor responsible for implementing a concomitant system of duties and obligations. Bentham, in other words, was pointing out that the mere declaration of a right doesn't mean that it is met; in his far more elegant terms, "wants are not means; hunger is not bread" (Bentham 2002: 330). I have also argued elsewhere that one of the most radical moments in the definition of food sovereignty is the layering of different jurisdictions over which rights can be exercised (Patel 2010). When the call is for, variously, nations, peoples, regions and states to craft their own agrarian policy, there is a concomitant call for spaces of sovereignty. Food sovereignty has its own geographies determined by specific histories and contours of resistance. To demand food sovereignty is to demand specific arrangements to govern territory and space. At the end of the day, the power of rights talk is that rights imply a particular burden on a specified entity: the state. In blowing apart the notion that the state has a paramount authority, by pointing to the multivalent hierarchies of power and control that exist within the world food system, food sovereignty paradoxically displaces one sovereign but remains silent about the others. To talk of a right to anything, after all, summons up a number of preconditions that food sovereignty, because of its radical character, undermines.

That there might be a class of people who were not covered by the territory of the state was a concern that troubled Arendt: hence her analytical (and personal) concern with refugees, with people stripped of nation-state membership and thus denied the ability to call on a state government's power to deliver and protect their rights. Yet, as Bentham suggests, talk of rights that exist simply because one is *human*, as Arendt argues for, is talk without substance. For who will guarantee the rights, for example, of those without a country? Which sovereign, for instance, guarantees the human rights of Palestinians, a people with a nation but no state?

Building on Arendt's work, Seyla Benhabib (2002) offers one of the more thoughtful Habermasian extensions of the idea of human rights, discussing the notion of a "right to rights" helpfully. Without repeating her arguments, she ultimately makes the case for a Kantian politics of cosmopolitan federalism and moral universalism (Benhabib 2004) in which different sovereignties and political jurisdictions layer atop one another, guided by fundamentally shared principles of rights. It is useful to see that the ideas of multiple "democratic attachments," which is Benhabib's way of referring to the many fealties that a single citizen can have (for example, Londoner, immigrant, British, European), can be attached to a longer tradition of political theory. But while expanding the conceptual resources available to discuss the existence of multiple and competing sovereignties, the Kantian call for cosmopolitan federalism and moral universalism looks very different under Benhabib's interpretation than advocates of food sovereignty might wish. For Benhabib, a good if imperfect working example of the kind of multiple and overlapping juridical sovereignties that are necessary to deal with the new political conjuncture is the European Union (Benhabib 2005). Within the E.U., a citizen can appeal to government at municipal, regional, national and Europe-wide levels, with each successive level trumping the ones below it. And, indeed, this looks like a very un-Westphalian system of rights provision. The cosmopolitan federalism element, with overlapping geographies over which one might claim rights, looks familiar in the definitions of food sovereignty.

But there's a problem. The E.U., despite its multifaceted sovereignties, is not a place characterized by food sovereignty. Indeed, its Common Agricultural Policy is the subject of scathing critique from within Europe by members of La Vía Campesina, and the E.U.'s Economic Partnership Arrangements violate the basic terms of food sovereignty in the Global South. This suggests that it is insufficient to consider only the structures that might guarantee the rights that constitute food sovereignty. It is also vital to consider the substantive policies, and politics, that go to make up food sovereignty. In other words, a simple appeal to rights talk cannot avoid tough questions around the substance and priority of those rights. In other words, while food sovereignty might be achieved through cosmopolitan federalism, if we're to understand what it looks like, we'll need also to look at the second part of Benhabib's dyad: moral universalism. Food sovereignty's multiple geographies have, despite their variety, a few core principles — and they are ones that derive from the politics through which La Vía Campesina was forged.

The Trace of Partial Universality in La Vía Campesina

The history of La Vía Campesina has been well documented elsewhere (Desmarais 2007), but one of the movement's central characteristics is the in-principle absence of a policymaking secretariat. Integral to the functioning of La Vía Campesina is the absence of a sovereign authority dictating what any member organization or country can do. As an organization forged in resistance to autocratic and unaccountable policymaking, largely carried out by the World Bank together with local elites, this suspicion of policies imposed from above is unsurprising within La Vía Campesina. Yet no organization can be a part of La Vía Campesina without subscribing to its principles, which are iterated at each international conference. These principles provide the preconditions for participation in La Vía Campesina's politics, and it's not surprising that the principles should find their analogue in the definition of food sovereignty. Another return to the definitions shows that there are a number of preconditions before food sovereignty can be achieved. Bear in mind, of course, that food sovereignty itself is a precondition for food security. Yet before any of this can be attained, there are a number of non-negotiable elements, preconditions if you will, for the preconditions for food security to exist.

The Nyéléni Declaration suggests that there is a range of conditions that are necessary for food sovereignty to obtain, such as living wages, tenure and housing security and cultural rights, as well as an end to the dumping of goods below the cost of production, disaster capitalism (Klein 2007), colonialism, imperialism and GMOs, in the service of a future where, among other things, "agrarian reform revitalises inter-dependence between consumers and producers" (La Vía Campesina 2007: 2). Specifically, these changes include a commitment to women's rights, not merely over property but over a full spectrum of social, physical and economic goods. It is worth noting that La Vía Campesina demands are for "access to land" rather than "ownership of land." The whole scope of power through ownership — be it of land, intellectual property rights or gene patents — is challenged by La Vía Campesina in various ways. This represents a fundamental challenge not only to neoliberal trade and policy strategies but also to the foundations of capitalism itself, insofar as it advances through enclosure and privatization.

It is here, I suggest, that we can use a feminist analysis to blow open an important set of questions around food sovereignty, specifically around the prioritization of rights. Under neoliberalism, as Monsalve evocatively suggests (2006), women's rights have become a Trojan Horse; the project of "giving rights to women" has been conscripted to spread a particular economic agenda founded on the primacy of individual private property rights. Other rights, such as those to education, health-care, social assistance and public investment derive, if at all, as rights secondary to individual private property. While women's rights to property are unarguably important, the attainment of the right cannot be understood as a sufficient means to "level the playing field for women." In a country with equal rights to property for all, the fact that some have more resources than others, and therefore are able to command more property than others, reflects underlying and persistent inequalities in power that make the ability to trade property a comparatively trivial right.

This base inequality in power is one that food sovereignty, sometimes explicitly, seeks to address. And it is here, in challenging deep inequalities of power, that I argue we see the core of food sovereignty. There is, at the heart of food sovereignty, a radical egalitarianism in the call for a multifaceted series of "democratic attach-ments." Claims around food sovereignty address the need for social change such that the capacity to shape food policy can be exercised at all appropriate levels. To make those rights substantive requires more than a sophisticated series of juridical sovereignties. To make the right to shape food policy meaningful is to require that *everyone* be able substantively to engage with them. But the prerequisites for this are a society in which the equality-distorting effects of sexism, patriarchy, racism and class power have been eradicated. Activities that instantiate this kind of radi-cal "moral universalism" are the necessary precursor to the formal "cosmopolitan federalism" that the language of rights summons. And it is by these activities that we shall know food sovereignty.

The canvas on which inequalities of power need to be tackled is vast. It might be argued that in taking this aggressively egalitarian view, I've opened up the project of food sovereignty so wide that it becomes everything and nothing. In my defence, I'd like to call on the Tanzanian political theorist, lawyer and activist Issa Shivji. In *Not Yet Democracy*, his brilliant analysis of land reform in Tanzania (1998), he addresses the question of what it will take for Tanzania to become a fully functioning democracy. He sees land reform as one of the central issues and argues forcefully that for the franchise to be meaningful, resources need to be dis-tributed as equally as the right to vote. In a poignant introduction to the book, he talks about how his daughters will grow up in a country that contains only the most cosmetic features of democracy and that their ability to be full and active citizens will be circumscribed because of the government's refusal to address the tough questions of resource distribution. Shivji's point is one that applies to the logic of food sovereignty, because both he and food sovereignty advocates are concerned, at the end of the day, with democracy. Egalitarianism, then, is not something that happens as a consequence of the politics of food sovereignty. It is a prerequisite to have the democratic conversation about food policy in the first place.

In taking this line, it looks like I'm violating the first rule of food sovereignty.

The genesis of the concept was designed precisely to prevent the kind of pinning down of interpretation that I attempt in this essay. But my interpretation doesn't pre-empt others, nor does it set in stone a particular political program. In making my interpretation, I am merely identifying and making explicit some of the commitments that are already implicit in the definition of food sovereignty. If we talk about food sovereignty, we talk about rights, and if we do that, we must talk about ways to ensure that those rights are met across a range of geographies, by everyone and in substantive and meaningful ways.

This isn't likely to be an interpretation that goes down agreeably among all stakeholders. In taking this egalitarianism seriously, several important social relations need to be addressed. La Vía Campesina has already identified the home as one such locus of social relations. What else can it mean when food sovereignty calls for women's rights to be respected than that the patriarchal traditions that characterize every household and every culture must, without exception, undergo transformation. The relations between farmers and farmworkers, too, are ones that are characterized by structural inequalities in power. Quite how La Vía Campesina members address this is not my place to say, and that's as well, because I'm very far from sure about the answer. But the fact that the question needs to be addressed is, in my mind, clear. Although the individual democratic movements within La Vía Campesina come at these issues from different starting points, traditions and politics, it seems to me that the questions about power, complicity and the profundity of a commitment to egalitarianism are ones that, by dint of their commitment to food sovereignty, the movements will ultimately have to address.

Notes

This chapter first appeared as "Food Sovereignty" in *Journal of Peasant Studies,* 36, 3 (2009). Reprinted with permission.

1.   See Michaels (2008), for instance, on the politics of "big tent" diversity being perfectly compatible with the neoliberal project.

References

Araghi, F. 1995. "Global Depeasantization, 1945–1990." *The Sociological Quarterly* 36, 2.

Arendt, H. 1967. *The Origins of Totalitarianism.* London: Allen & Unwin.

Bakan, J. 2004. *The Corporation: The Pathological Pursuit of Profit and Power.* New York: New Press.

Balakrishnan, R., and D. Elson. 2008. "Auditing Economic Policy in the Light of Obligations on Economic and Social Rights." *Essex Human Rights Review* 5, 1.

Benhabib, S. 2002. "Political Geographies in a Global World: Arendtian Reflections." *Social Research* 69, 2.

_____. 2004. *The Rights of Others: Aliens, Residents and Citizens.* Cambridge: Cambridge University Press.

_____. 2005. "Borders, Boundaries, and Citizenship." *PS: Political Science and Politics* 38, 4.

Bentham, J. 2002. *Rights, Representation and Reform: Nonsense Upon Stilts and Other Writings on the French Revolution.* Oxford: Oxford University Press.

Desmarais, A.A. 2007. *La Vía Campesina: Globalization and the Power of Peasants.* Halifax: Fernwood Publishing; London, Ann Arbor; MI: Pluto Press.

FAO (Food and Agricultural Organization of the United Nations). 2001. *The State of Food Insecurity in the World 2001.* Rome.

_____. 2003. "Trade Reforms and Food Security: Conceptualising the Linkages." Rome: Commodity Policy and Projections Service, Commodities and Trade Division.

Klein, N. 2007. *The Shock Doctrine: The Rise of Disaster Capitalism.* New York: Metropolitan Books/Henry Holt.

La Vía Campesina. 1996. "The Right to Produce and Access to Land." Available at <voiceoftheturtle.org/library/1996%20Declaration%20of%20Food%20Sovereignty.pdf.>.

_____. 2007. "Nyéléni Declaration." Sélingué, Mali: Forum for Food Sovereignty.

McMichael, P. 2008. "Food Sovereignty, Social Reproduction, and the Agrarian Question." In A.H. Akram-Lodhi and C. Kay (eds.), *Peasant Livelihoods, Rural Transformation and the Agrarian Question.* London: Routledge.

Michaels, Walter Benn. 2008. "Against Diversity." *New Left Review* 52 (July, Aug).

Monsalve, S. 2006. "Gender and Land." In M. Courville, R. Patel and P. Rosset (eds.), *Promised Land: Competing Visions of Agrarian Reform.* Oakland, CA: Food First Books.

Patel, R. 2006. "Transgressing Rights: La Vía Campesina's Call for Food Sovereignty." *Feminist Economics* 12, 4.

_____. 2010. *The Value of Nothing: How To Reshape Market Society and Reclaim Democracy.* New York: Picador.

Peoples' Food Sovereignty Network. 2002. "Statement on Peoples' Food Sovereignty."

Rajagopal, B. 2000. "From Resistance to Renewal: The Third World, Social Movements, and the Expansion of International Institutions." *Harvard International Law Journal* 41, 2.

Rosset, P. 2003. "Food Sovereignty: Global Rallying Cry of Farmer Movements." Oakland, CA: Institute for Food and Development Policy.

Shivji, I.G. 1998. *Not Yet Democracy: Reforming Land Tenure in Tanzania.* London: IIED.

United Nations. 1975. "Report of the World Food Conference, Rome November 5–16, 1974." New York: United Nations.

Windfuhr, M., and J. Jonsén. 2005. *Food Sovereignty: Towards Democracy in Localized Food Systems.* Rugby, Warwickshire: ITDG Publishing.

Appendix 1
The Right to Produce and Access to Land

Food Sovereignty: A Future without Hunger

We, the Vía Campesina, a growing movement of farm workers, peasant, farm and indigenous peoples' organizations ... know that food security cannot be achieved without taking full account of those who produce food. Any discussion that ignores our contribution will fail to eradicate poverty and hunger.

Food is a basic human right. This right can only be realized in a system where food sovereignty is guaranteed. Food sovereignty is the right of each nation to maintain and develop its own capacity to produce its basic foods respecting cultural and productive diversity. We have the right to produce our own food in our own territory. Food sovereignty is a precondition to genuine food security.

...

We are determined to create rural economies which are based on respect for ourselves and the earth, on food sovereignty and fair trade. Women play a central role in household and community food sovereignty. Hence they have an inherent right to resources for food production, land, credit, capital, technology, education and social services, and equal opportunity to develop and employ their skills. We are convinced that the global problem of food insecurity can and must be resolved. Food sovereignty can only be achieved through solidarity and the political will to implement alternatives.

Long-term food security depends on those who produce food and care for the natural environment. As the stewards of food producing resources we hold the following principles as the necessary foundation for achieving food security.

Food — a Basic Human Right

Food is a basic human right. Everyone must have access to safe, nutritious and cultural appropriate food in sufficient quantity and quality to sustain a healthy life with full human dignity. Each nation should declare that access to food is a constitutional right and guarantee the development of the primary sector to ensure the concrete realization of this fundamental right.

Agrarian Reform

[Food sovereignty] demands genuine agrarian reform which gives landless and farming people — especially women — ownership and control of the land they work and returns territories to Indigenous peoples. The right to land must be free of discrimination on the basis of gender, religion, race, social class or ideology; land belongs to those who work it. Peasant families, especially women, must have access to productive land, credit, technology, markets and extension services.

Governments must establish and support decentralized rural credit systems that prioritize the production of food for domestic consumption to ensure food sovereignty. Production capacity rather than land should be used as security to guarantee credit. To encourage young people to remain in rural communities as productive citizens, the work of producing food and caring for the land has to be sufficiently valued both economically and socially. Governments must make long-term investments of public resources in the development of socially and ecologically appropriate rural infrastructure.

Protecting Natural Resources

Food sovereignty entails the sustainable care and use of natural resources especially land, water and seeds. We, who work the land, must have the right to practice sustainable management of natural resources and to preserve biological diversity. This can only be done from a sound economic basis with security of tenure, healthy soils and reduced use of agro-chemicals. Long-term sustainability demands a shift away from dependence on chemical inputs, on cash-crop monocultures and intensive, industrialized production models. Balanced and diversified natural systems are required. Genetic resources are the result of millennia of evolution and belong to all of humanity. They represent the careful work and knowledge of many generations of rural and indigenous peoples. The patenting and commercialization of genetic resources by private companies must be prohibited. The World Trade Organization's Intellectual Property Rights Agreement is unacceptable. Farming communities have the right to freely use and protect the diverse genetic resources, including seeds, which have been developed by them throughout history.

Reorganizing the Food Trade

Food is first and foremost a source of nutrition and only secondarily an item of trade. National agricultural policies must prioritize production for domestic consumption and food self-sufficiency. Food imports must not displace local production nor depress prices. This means that export dumping or subsidized export must cease. Peasant farmers have the right to produce essential food staples for their countries and to control the marketing of their products. Food prices in domestic and international markets must be regulated and reflect the true cost of producing that food. This would ensure that peasant families have adequate incomes. It is unacceptable that the trade in foodstuffs continues to be based on the economic exploitation of the most vulnerable — the lowest earning producers — and the further degradation of the environment. It is equally unacceptable that trade and production decisions are increasingly dictated by the need for foreign currency to meet high debt loads. These debts place a disproportionate burden on rural peoples.... [T]hese debts must be forgiven.

Ending the Globalization of Hunger

Food sovereignty is undermined by multilateral institutions and by speculative capital. The growing control of multinational corporations over agricultural policies has been facilitated by the economic policies of multilateral organizations such as the WTO, World Bank and the IMF.... [T]he regulation and taxation of speculative capital and a strictly enforced Code of Conduct for transnational corporations [are necessary].

Social Peace – A Pre-Requisite to Food Sovereignty

Everyone has the right to be free from violence. Food must not be used as a weapon. Increasing levels of poverty and marginalization in the countryside, along with the growing oppression of ethnic minorities and indigenous populations aggravate situations of injustice and hopelessness. The increasing incidence of racism in the countryside, ongoing displacement, forced urbanization and repression of peasants cannot be tolerated.

Democratic Control

Peasants and small farmers must have direct input into formulating agricultural policies at all levels. This includes the current FAO World Food Summit from which we have been excluded. The United Nations and related organizations will have to undergo a process of democratization to enable this to become a reality. Everyone has the right to honest, accurate information and open and democratic decision-making. These rights form the basis of good governance, accountability and equal participation in economic, political and social life, free from all forms of discrimination. Rural women, in particular, must be granted direct and active decision-making on food and rural issues.

Source: Adapted from La Vía Campesina, November 11–17, 1996, Rome Italy, available at <www.viacampesina.org>.

Appendix 2

The statement Priority to Peoples' Food Sovereignty was launched on November 6, 2001, by a group of organizations involved in the Our World Is Not for Sale Coalition, just prior to the Fourth Ministerial Conference of the World Trade Organization (WTO) in Doha. The statement was prepared by La Vía Campesina, COASAD, Collectif Stratégies Alimentaires, ETC Group, Focus on the Global South, Foodfirst/Institute for Food and Development Policy, Friends of the Earth Latin America and Caribbean, Friends of the Earth England, Wales and Northern Ireland, GRAIN, Institute for Agriculture and Trade Policy, IBON Foundation, and Public Citizen's Energy and Environment Program. Since the statement was launched numerous movements have signed on.

Our World Is Not For Sale

Priority to Peoples' Food Sovereignty

WTO out of Food and Agriculture

Agriculture and food are fundamental to all people, in terms of both production and availability of sufficient quantities of safe and healthy food, and as foundations of healthy communities, cultures and environments. All of these are being undermined by the increasing emphasis on neo-liberal economic policies promoted by leading political and economic powers such as the United States (US) and the European Union (EU), and realised through global institutions such as the World Trade Organisation (WTO), International Monetary Fund (IMF) and the World Bank (WB). Instead of securing food for the people of the world, these institutions have presided over a system that has prioritised export-oriented production, increased global hunger and malnutrition, and alienated millions of people from productive assets and resources such as land, water, seeds, technology and know-how. Fundamental change to this global regime is urgently required.

Peoples' Food Sovereignty is a Right

In order to guarantee the independence and food sovereignty of all of the world's peoples, it is essential that food be produced though diversified, farmer-based production systems. Food sovereignty is the right of peoples to define their own agriculture and food policies, to protect and regulate domestic agricultural production and trade in order to achieve sustainable development objectives, to determine the extent to which they want to be self reliant, and to restrict the dumping of products in their markets. Food sovereignty does not negate trade, but rather, it promotes the formulation of trade policies and practices that serve the rights of peoples to safe, healthy and ecologically sustainable production.

Governments must uphold the rights of all peoples to food sovereignty and security, and adopt policies that promote sustainable, family-farm based production rather than industry-led, high-input and export oriented production. This in turn demands that they put in place the following measures:

I.   Market Policies
-   Ensure adequate remunerative prices for all farmers;
-   Exercise the rights to protect domestic markets from imports at low prices;
-   Regulate production on the internal market in order to avoid the creation of surpluses;
-   Abolish all direct and indirect export supports;
-   Phase out domestic production subsidies that promote unsustainable agriculture and inequitable land tenure patterns, and target support at integrated agrarian reform programs as well as sustainable farming practices.

II.   Food Safety, Quality and the Environment
-   Adequately control the spread of diseases and pests while at the same time ensuring food safety;
-   Ban the use of dangerous technologies such as food irradiation, which lower the nutritional value of food and create toxins in food;
-   Establish food quality criteria appropriate to the preferences and needs of the people;
-   Establish national mechanisms for quality control of all food products so that they comply with high environmental, social and health quality standards;
-   Ensure that all food inspection functions are performed by appropriate and independent government bodies, and not by private corporations or contractors.

III.   Access to Productive Resources
-   Recognise and enforce communities' legal and customary rights to make decisions concerning their local, traditional resources, even where no legal rights have previously been allocated;
-   Ensure equitable access to land, seeds, water, credit and other productive resources;
-   Prohibit all forms of patenting of life or any of its components, and the appropriation of knowledge associated with food and agriculture through intellectual property rights regimes;
-   Protect farmers', indigenous peoples' and local community rights over plant genetic resources and associated knowledge — including farmers' rights to exchange and reproduce seeds.

IV.   Production–Consumption
-   Develop local food economies based on local production and processing, and the development of local food outlets.

V.   Genetically Modified Organisms (GMOs)

- Ban the production of, and trade in genetically modified (GM) seeds, foods, animal feeds and related products.
- Encourage and promote alternative agriculture and organic farming based on indigenous knowledge and sustainable agriculture practices.
- Expose and actively oppose the various methods (direct and indirect) by which agri-business corporations such as Monsanto, Syngenta, Aventis/Bayer and DuPont are bringing GM crop varieties into agricultural systems and environments.

V.   Transparency of Information and Corporate Accountability

- Provide clear and accurate labelling of food and feed-stuff products based on consumers' and farmers' rights to access to information about content and origins;
- Establish binding regulations on all companies to ensure transparency, accountability and respect for human rights and environmental standards;
- Establish anti-trust laws to prevent the development of industrial monopolies in the food and agricultural sectors.
- Hold corporate entities and their directors legally liable for corporate breaches of environmental and social laws, and of national and international laws and agreements.

Trade Rules Must Guarantee Food Sovereignty

Global trade must not be afforded primacy over local and national developmental, social, environmental and cultural goals. Priority should be given to affordable, safe, healthy and good quality food, and to culturally appropriate subsistence production for domestic, sub-regional and regional markets. Current modes of trade liberalisation, which allows market forces and powerful transnational corporations (TNCs) to determine what and how is produced, and how food is traded and marketed, cannot fulfil these crucial goals.

"No" to Neo-liberal Policies in Food and Agriculture

The undersigned denounce the 'liberalisation' of farm product exchanges as promoted through bilateral and regional free trade agreements, and multilateral institutions such as the IMF, the World Bank and the WTO. We condemn the dumping of agricultural products in all markets, and especially in Third World countries where it has severely undermined domestic production. Neo-liberal policies coerce countries into specialising in agricultural production in which they have a so-called "comparative advantage" and then trading along the same lines. However, export orientated production is being pushed at the expense of domestic food production, and production means and resources are increasingly controlled by large transnational corporations.

Rich governments continue to heavily subsidise export oriented agricultural production in their countries, with the bulk of support going to large producers. The majority of tax-payers' funds are handed out to big business — large producers, traders and retailers — who engage in unsustainable agricultural and trading practices, and not to small-scale family producers who produce much of the food for the internal market, often in more sustainable ways.

These export-oriented policies have resulted in market prices for commodities that are far lower than their real costs of production. This has encouraged and perpetuated dumping, and provided TNCs with opportunities to buy cheap agricultural products which are then sold at significantly higher prices to consumers in both the North and the South. The larger parts of important agricultural subsidies in rich countries are in fact subsidies for agro-industry, traders, retailers and a minority of the largest producers.

The adverse effects of these policies and practices are becoming clearer every day. They lead to the disappearance of small-scale, family farms in both the North and South; poverty has increased, especially in the rural areas; soils and water have been polluted and degraded; biological diversity has been lost; and natural habitats destroyed.

Dumping

Dumping occurs when goods are sold at less than their cost of production. This can be the result of subsidies, structural distortions such as monopoly control over markets and distribution, and also the inability of current economic policy to factor in externalities such as the depletion of water and soil nutrients and pollution resulting from industrial agricultural methods. Dumping under the current neo-liberal policies is conducted in North-South, South-North and South-South and North-North trade. Whatever the form, dumping ruins small-scale local producers in both the countries of origin and sale.

For example:
-   Imports by India of dairy surpluses subsidized by the European Union, which had negative impacts on local, family based dairy production;
-   Exports of industrial pork from the USA to the Caribbean, which proved ruinous to Caribbean producers;
-   Imports by Ivory Coast of European pork at subsidised prices which are three times lower than the production costs in Ivory Coast;
-   Chinese exports of silk threads to India at prices far lower than the costs of production in India; this has been seriously damaging for hundreds of thousands of farmer families in Southern India;
-   On one hand the import of cheap maize from the US to Mexico — the centre of the origin of maize--ruins Mexican producers; on the other hand the export of vegetables at low prices from Mexico to Canada ruin producers in Canada.

Dumping practises must to be stopped. Countries must be able to protect their home markets against dumping and other trade practices that prove damaging to local producers. Exporting countries must not be allowed to dump surpluses on the international market, and should respond to real demands for agricultural goods and products in ways that do not undermine domestic production.

There is no "World Market" of Agricultural Products

The so called "world market" of agricultural products does not exist. What exists is, above all, an international trade of surpluses of milk, cereals and meat dumped primarily by the EU, the US, and other members of the CAIRNS group. Behind the faces of national trade negotiators are powerful TNCs such as Monsanto and Cargill who are the real beneficiaries of domestic subsidies and supports, international trade negotiations and the global manipulations of trade regimes. At present, international trade in agricultural products involves only ten percent of total worldwide agricultural production and is mainly an exchange between TNCs from the US, EU and a few other industrialised countries. The so called "world market price" is extremely unstable and has no relation with the costs of production. It is far too low because of dumping, and therefore, it is not an appropriate or desirable reference for agricultural production.

The WTO Dismisses Calls for Reform

The WTO is undemocratic and unaccountable, has increased global inequality and insecurity, promotes unsustainable production and consumption patterns, erodes diversity, and undermines social and environmental priorities. It has proven impervious to criticisms regarding its work and has dismissed all calls for reform. Despite promises to improve the system made at the Seattle Minesterial Meeting in 1999, governance in the WTO has actually become worse. Rather than addressing existing inequities and power imbalances between rich and poor countries, the lobby of the rich and powerful in the WTO is attempting to expand the WTO's mandate to new areas such as environment, labour, investment, competition and government procurement.

The WTO is an entirely inappropriate institution to address issues of agriculture and food. The undersigned do not believe that the WTO will engage in profound reform in order to make itself responsive to the rights and needs of ordinary people. Therefore, the undersigned are calling for all food and agricultural concerns to be taken out of WTO jurisdiction through the dismantling of the Agreement on Agriculture (AoA) and removing or amending the relevant clauses on other WTO agreements so as to ensure the full exclusion of food and agriculture from the WTO regime. These include: the Agreement on Trade Related Intellectual Property Rights (TRIPs), Sanitary and Phytosanitary measures (SPS), Technical Barriers to Trade (TBT), Quantitative Restrictions (QRs), Subsidies and Countervailing Measures (SCM), and the General Agreement on Trade in Services (GATS).

A Role for Trade Rules in Agricultural and Food Policies?

Trade in food can play a positive role, for example, in times of regional food in-security, or in the case of products that can only be grown in certain parts of the world, or for the exchange of quality products. However, trade rules must respect the precautionary principle to policies at all levels, recognise democratic and participatory decision-making, and place peoples' food sovereignty before the imperatives of international trade.

An Alternative Framework

To compliment the role of local and national governments, there is clear need for a new and alternative international framework for multilateral regulation on the sustainable production and trade of food and other agricultural goods. Within this framework, the following principles must be respected:

1. Peoples' food sovereignty;
2. The right of all countries to protect their domestic markets by regulating all imports which undermine their food sovereignty;
3. Trade rules that support and guarantee food sovereignty;
4. Upholding gender equity and equality in all policies and practices concerning food production;
5. The precautionary principle;
6. The right to information about the origin and content of food items;
7. Genuine international democratic participation mechanisms;
8. Priority to domestic food production, sustainable farming practices and equitable access to all resources;
9. Support for small farmers and producers to own, and have sufficient control over means of food production;
10. An effective ban on all forms of dumping in order to protect domestic food production; this would include supply management by exporting countries to avoid surpluses and the rights of importing countries to protect internal markets against imports at low prices;
11. Prohibition of biopiracy and patents on living matter — animals, plants, the human body and other life forms — and any of its components, including the development of sterile varieties through genetic engineering;
12. Respect for all human rights conventions and related multilateral agreements under independent international jurisdiction.

The undersigned affirm the demands made in other civil-society statements such as Our World is Not for Sale: WTO-Shrink or Sink, and Stop the GATS Attack Now. We urge governments to immediately take the following steps:

1. Cease negotiations to initiate a new round of trade liberalisation and

halt discussions to bring 'new issues' into the WTO. This includes further discussions on such issues as investment, competition, government procurement, biotechnology, services, labour and environment;

2.   Cancel further trade liberalisation negotiations on the WTO's AoA through the WTO's built-in agenda;

3.   Cancel the obligation of accepting the minimum importation of 5% of internal consumption; all compulsory market access clauses must similarly be cancelled immediately;

4.   Undertake a thorough review of both the implementation, and the environmental and social impacts of existing trade rules and agreements (and the WTO's role in this system) in relation to food and agriculture;

5.   Initiate measures to remove food and agriculture from under the control of the WTO through the dismantling of the AoA, and through the removal or amendment of relevant clauses in the TRIPs, GATS, SPS, TBT and SCM agreements; and further, by replacing these with a new Convention on Food Sovereignty and Trade in Food and Agriculture;

6.   Revise intellectual property policies to prohibit the patenting of living matter and any of their components and limit patent protections in order to protect public health and public safety;

7.   Halt all negotiations on GATS, and dismantle the principle of "progressive liberalisation" in order to protect social services and the public interest;

8.   Implement genuine agrarian reform and ensure the rights of peasants to crucial assets such as land, seed, water and other resources;

9.   Initiate discussions on an alternative international framework on the sustainable production and trade of food and agricultural goods. This framework should include:

   - A reformed and strengthened United Nations (UN), active and committed to protecting the fundamental rights of all peoples, as being the appropriate forum to develop and negotiate rules for sustainable production and fair trade;
   - An independent dispute settlement mechanism integrated within an international Court of Justice, especially to prevent dumping;
   - A World Commission on Sustainable Agriculture and Food Sovereignty established to undertake a comprehensive assessment of the impacts of trade liberalisation on food sovereignty and security, and develop proposals for change; these would include the agreements and rules within the WTO and other regional and international trade regimes, and the economic policies promoted by International Financial Institutions and Multliateral Development Banks; such a commission could be constituted of and directed by representatives from various social and cultural groups, peoples' movements, professional fields, democratically elected representatives and appropriate

multilateral institutions;

- An international, legally binding Treaty that defines the rights of peasants and small producers to the assets, resources and legal protections they need to be able to exercise their right to produce; such a treaty could be framed within the UN Human Rights framework, and linked to already existing relevant UN conventions;
- An International Convention that replaces the current Agreement on Agriculture (AoA) and relevant clauses from other WTO agreements and implements within the international policy frame work the concept of food sovereignty and the basic human rights of all peoples to safe and healthy food, decent and full rural employment, labour rights and protection, and a healthy, rich and diverse natural environment and incorporate trading rules on food and agriculture commodities.

A Broad Alliance with an Agenda for Change!

The impacts of the neo-liberal policies are all too evident and increasingly understood and challenged by civil society across the world. The pressure for change is increasing.

In the run up to the next WTO Ministerial Meeting and in the coming years, the undersigned will continue to reveal the adverse effects of neo-liberal trade and economic policies on agriculture and food, and to propose alternatives to the current global trade regime.

This declaration is a clear sign of the determination that unites social movements and other civil-society actors world-wide in their struggle to democratise international policies, and to work towards institutions that are capable of embracing and defending sustainable approaches to food and agriculture.

Index